THE LAW OF STORMS

THE TRUE STORY OF THE RMS RHONE AND THE GREAT VIRGIN ISLANDS HURRICANE OF 1867

JOAN MARTELLI

ISBN 978-90-827838-1-0

Paperback

First photo: Believed to be the only existing photograph of the *Rhone.* Also possibly *Rhone*'s identical sister ship, *Douro*. (1865). COURTESY: NATIONAL MARITIME MUSEUM, GREENWICH, LONDON

Cover: Photo taken from Tortola looking over Sir Francis Drake Channel toward Salt Island BVI.

Author photo: credit: Neil Rubino

Maps: Google

For my parents, John and Bernadette Martelli

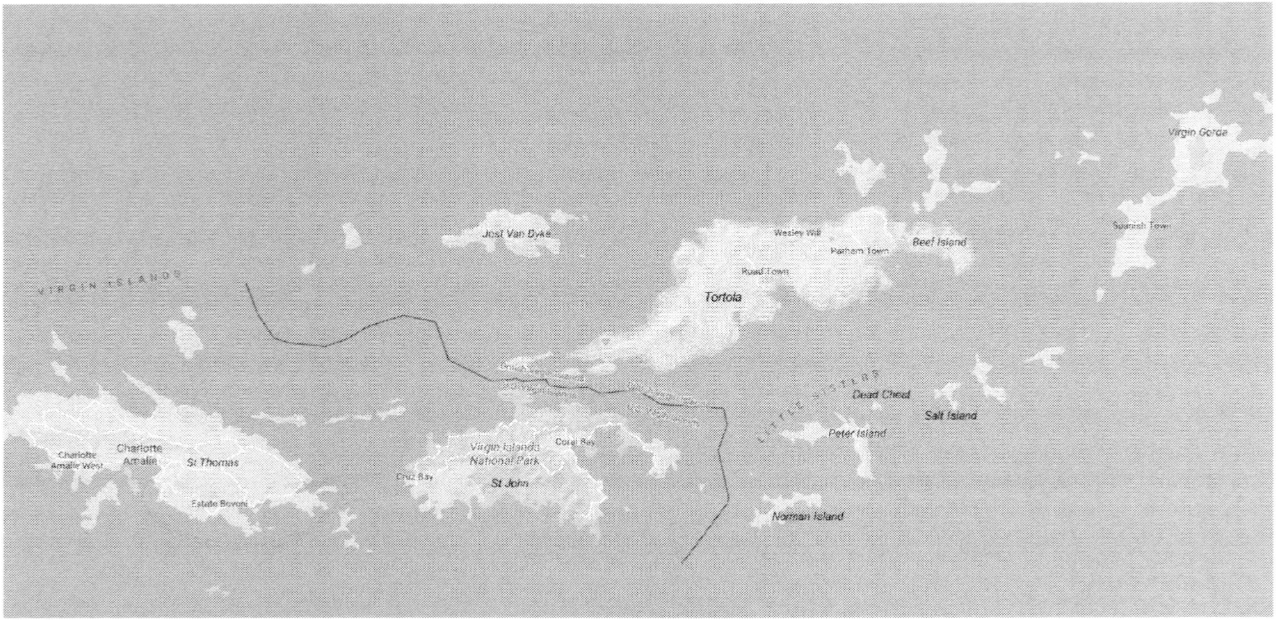
Virgin Gorda
Spanish Town
Jost Van Dyke
Wesley Will
Parham Town
Beef Island
Road Town
Tortola
VIRGIN ISLANDS
LITTLE SISTERS
Dead Chest
Salt Island
Peter Island
Norman Island
Coral Bay
Virgin Islands National Park
St John
Cruz Bay
Charlotte Amalie West
Charlotte Amalie
St Thomas
Estate Bovoni

CONTENTS

	The Law of Storms	vii
	Introduction	ix
1.	Maiden Voyage	1
2.	Law of Storms	16
3.	Dancing on the Sea	30
4.	Captain Frederick Woolley	49
5.	Black Jack Island	62
6.	Safe Haven	68
7.	The Hurricane Approaches	81
8.	The Eye	90
9.	Drifting Men	105
10.	A Riot of Wrecks	117
11.	Sad Intelligence	121
12.	Tsunami	137
13.	Lucky and Good	155
14.	Diving the Rhone	168
	Afterword	175
	Acknowledgments	179
	About the Author	183

THE LAW OF STORMS

From the archives of the National Maritime Museum, Greenwich, London

INTRODUCTION

October 29, 2017

When I began writing *The Law of Storms,* the US and British Virgin Islands looked very different than they do today. The BVI, in particular, was a dreamlike haven of lush vegetation, superb sailing, scuba diving and charming beach resorts.

Then came Hurricane Irma. On September 6, 2017, the Category 5 storm with wind speeds of more than 150 miles an hour, decimated the Virgin Islands. The eye of the monster actually passed directly over Tortola and Virgin Gorda, the two biggest islands in the BVI. The scorching winds stripped bare virtually every tree, wrecked boats, tore the roofs off hotels and businesses and demolished the houses of thousands of local islanders many of whom (even as they were interviewed in the ruins of their homes) vowed to rebuild because, after all, they lived in "paradise." One month later the hillsides in Tortola were starting to turn green again, ferries were running, restaurants were reopening and businesses of every kind were optimistic about being able to welcome tourists back. But the damage estimate is in the billions and it will take a long time to rebuild.

Although the scale of the destruction from Hurricane Irma was unprecedented, the event—a direct hit by a major hurricane—was not.

Exactly 150 years before Irma struck the Caribbean, the eye of a Category 3 hurricane passed directly over the British and US Virgin Islands, devastating homes, lives and landscape alike—and sinking what was considered at the time one of the most modern and luxurious ships in the world, the Royal Mail Steamship *Rhone.*

Most of the passengers and crew of the RMS *Rhone* died in that storm but some managed to survive and the legend of the *Rhone* lived on.

Today the colossal ruin of the 310-foot iron-hulled vessel, which some call the *Titanic* of her time, lies in shallow water not far from Tortola. The wreck of the *Rhone* is a protected national marine park, ranked as one of the best dive sites in the Caribbean.

The Law of Storms is a work of non-fiction that reads more like a novel. It recounts the true story of the *Rhone* from her maiden voyage in 1865 to her last on October 29, 1867. The book explores the personal and professional lives of her two rival captains and the social and political forces at play in the Virgin Islands as they collided with unstoppable powers of nature which, in the 1860s, included not only a hurricane but a yellow fever outbreak, an earthquake and a tsunami. It also details how each of the 25 people who escaped the wreck of the *Rhone* managed to survive.

Every shipwreck has a story. The story of the *Rhone* as told in *The Law of Storms* is a bridge between the dramatic events of the past and the reality of today. Undoubtedly, as they begin to rebuild and recover, the true story of the *Rhone* will be part of the Virgin Islands' future as well.

It was a time long ago but not unlike our own—a time when one storm could change everything.

1

MAIDEN VOYAGE

October 9, 1865
Southampton, England

Standing on the deck of the RMS *Rhone* on departure day, Captain Robert Woolward must have considered himself quite fortunate.

Just 39-years old, Woolward had already commanded ten ships belonging to the prestigious Royal Mail Steam Packet Company. Not all of those ships had been to his liking. Ugly, old and like a "Chinese junk" are among the unflattering words he used to describe some of them.

But this was different.

The *Rhone* was the most beautiful ship Robert Woolward had ever seen and he was about to take her across the Atlantic on her maiden voyage.

"The *Rhone* was to all intents and purposes a steam yacht fitted with every modern improvement and convenience," Woolward said. He was very proud to be her commander.

Departure day was an especially busy one in the port of Southampton, where a parade of ships could regularly be seen arriving from and departing to places like New York, Rio, India and Australia. But even in this bustle of activity the *Rhone* stood out. She was special—a Victorian masterpiece.

She was big—three hundred and ten feet long—about the length of a European soccer or American football field. Between her two giant masts (each fitted with a full suit of sails) stood a stout black smoke funnel, the unmistakable mark of a modern steamship. Despite her size, the *Rhone* also appeared quite elegant thanks to her gracefully arching clipper ship bow.

The *Rhone* had four decks. The top two included saloons, a galley and passenger cabins with enough room to accommodate 253 first-class, 30 second-class and 30 third-class passengers.

The *Rhone*'s first-class saloon, where gourmet meals were served daily, was especially large, spanning the full 40-foot width of the ship. Decorated in the high style of the time, the room was apt to be laid out with paneling made of heavy oak, gilded pillars and cornices, solid walnut furniture, large patterned carpets and deep padded dimpled couches with skirts and dusters. Reporters described it as "splendid" and "chastely" decorated.

A few weeks earlier the *Rhone*'s saloon had been the setting for a grand event, a dinner with guests that included two notable women of the day. One was Queen Emma of the Sandwich Islands (Hawaii), respected in England for championing Anglican churches, schools and hospitals in her native land. The other was Lady Jane Franklin, widow of the famed Arctic explorer Sir John Franklin and a humanitarian in her own right. Reporters judged the banquette a great success and predicted the *Rhone* would live up to expectations as "one of the fastest and best vessels in the company's service."

Not just large and luxurious, the *Rhone* was also fast. Her state-of-the-art 500-horse power engine, powered by steam under pressure, set in motion a massive screw propeller whose rotating force could send the 2,738 gross registered ton *Rhone* racing through the ocean. At a time when the Royal Mail fleet averaged 10 knots, the *Rhone,* at her official time trial, clocked an average of about 14 and on one leg hit more than 15 knots.

A sizable crew was needed to operate such a ship. The *Rhone* had a crew of approximately 130 including officers, 30 seamen, 15 firemen, 15 coal trimmers, 12 waiters, five stewards, three cooks, two butchers, a baker, a barman and other deck hands as young as 15-years old.

And there was the captain, the single most important individual on the ship, a man on whose judgment and skill the lives of every crew member and passenger depended. The *Rhone's* captain, Robert Woolward, was extraordinary in his own right. Described in company reports as attentive to his duties, a steady sailor and an excellent navigator, Woolward was respected for his seamanship by both his superiors and crew. But there was another side to the captain, a mischievous one.

A fellow officer put it this way. While Woolward would sign letters to his employer—*Gentleman, I am your obedient, humble servant*—he was neither.

Robert Woolward was born—or as he puts it, "launched"—on April 15, 1826 in the coastal town of Ramsgate, England.

His father, John Cornelius Woolward, had served with the legendary Admiral Horatio Lord Nelson at Aboukir Bay where the English famously defeated Napoleon in the Battle of the Nile. When the fighting was over, Robert's father was left partially deaf. In compensation for his injury and his service, John Woolward was awarded the position of Harbor Master of Ramsgate, an important post that he held for more than 20 years.

But from an early age, it was clear Robert didn't fit the Royal Navy mold. He was, in his own words, "a bundle of mischief." By age nine he was sent off by his exasperated father to the Chatham House Academy where he benefited from good food and long school days that started at 5:40 a.m. and didn't end until after prayers at 8:20 at night.

By age 14 Woolward, like many boys in those days, left school and began work as an apprentice. On a ship called the *Diadem*, bound for Quebec, young Woolward had morning watch (4 a.m. to 6 a.m.) and evening watch (6 p.m. to 8 p.m.) and in between he worked as the ship's cook and steward. But the routine of hard work suited him and he thrived during his early training. He managed his duties well and proudly pointed out that although he was the youngest apprentice on the *Diadem,* he was the first to master the skill of steering her.

By age 15 Woolward was off to Asia. A family friend had helped secure him a position on *La Belle Alliance*, an East India Company merchant ship commissioned to carry supplies and troops for the "China Expedition," also known as the First Opium War. Woolward would spend more than three years visiting faraway ports like Hong Kong, Singapore and Bombay and in all that time *La Belle Alliance* sailed virtually accident free. A "lucky ship" Woolward called her. In the summer of 1844 she returned to London with a cargo of antimony ore, pepper and rhubarb. A now seasoned 18-year old sailor, Woolward was ready for his next adventure.

But finding another job wouldn't be easy. His many

applications resulted in two equally unhappy responses. Either he was too young and underqualified, or he was perfect for the job but it had just been filled.

"This state of things went on for some time," says Woolward, "and I was beginning to think I was either born a day too late, or that it would have been better if I had not been born at all."

Then on what he calls the "Right Day," Woolward walked into the offices of the Royal Mail Steam Packet Company on Moorgate Street in London and managed to get a meeting with its director and future Chairman, Captain Charles Mangles. The shipping company was a powerhouse at the time and, in the opinion of some historians, second to none in the world. Five years earlier the Royal Mail Steam Packet Company had signed a lucrative contract with the British government to provide mail service to the West Indies, a deal which bestowed a rich annual subsidy and demanded dependable and prompt service in return.

Royal Mail ships would carry news, correspondence and also millions of pounds worth of gold, silver and exotic goods like walrus ivory, whalebone, silkworm eggs, cigars, tobacco and even live turtles across the sea. They were also the first choice of well-to-do passengers including barons, ministers and princes.

Royal Mail steamships left Southampton like clockwork on the 2nd and 17th of each month, traveling to the islands of the Caribbean West Indies. By 1850 the company would expand its reach to include South America.

Although its head office was in London, the company's workaday home was a four story building across from the docks on #7-8 Canute Road in Southampton. The up-and-coming company and the historic port town seemed made for each other.

Southampton inns, shops and restaurants benefitted from the steady flow of travelers and the Royal Mail Steam Packet Company employed hundreds of local workers. Most of the port jobs, however, did not pay well and were not year-round which left dockworkers and their families living in poverty with three generations often crammed together in tight quarters.

But when it came to building the original Royal Mail fleet, no cost was spared. The company ordered fourteen new expensive steamships, each constructed to the same general specifications. They were approximately 1800 tons, paddlewheel propelled and oak hulled with expensive copper bottoms to protect them from the tropical torpedo worm. They carried two six-pound signal cannons (which would be fired to announce the arrival and departure from port with the mail) and some 30 muskets, pistols, cutlasses and pikes. These merchant ships were expected to be ready and able to fight if called into service by the Royal Navy.

Spectacular celebrations accompanied the launch of the first fourteen ships. A particularly large crowd of about 60,000 people gathered to witness the christening of the *Forth*. Reporters described the late afternoon event as having a fair-like quality with booths, flags and fireworks. People on foot, horseback, carriages and riding in two-wheeled tea caddies all made their way dockside for the launch. When the time came, Miss Colville (daughter of the Deputy Chairman of the Royal Mail Steam Packet Company) let go a small decanter of wine decorated with colored ribbons suspended by a cord which shattered against the side of the *Forth*.

“Cut the trigger!” came the call, and props holding the

great ship at bay were knocked away. Cannons fired. The crowd cheered, "Hurrah!" Carpenters who had helped build the ship, threw their hammers after her for luck and the *Forth* slipped seamlessly into the sea.

Unfortunately, all the excitement surrounding those first Royal Mail steamships was followed by an embarrassing series of disasters. Within a year, three of its celebrated new ships were lost. First the *Medina* ran up on a reef at Turks Island in the middle of the night. All the passengers and mail were saved but the ship was a total wreck. Then the *Isis* went aground in Puerto Rico, an accident that claimed the life of a ship's apprentice who fell overboard. Much more devastating was an incident involving the *Solway* which also ran afoul of a reef. In that case, 17 passengers and 18 officers and crew, including the captain, died. As Royal Mail historian T.A. Bushell puts it—"It could never be claimed that the Company was fortunate in its early days."

But for the young and ambitious Robert Woolward, the Royal Mail Steam Packet Company offered exciting new opportunities. And besides, he needed a job. So when Captain Mangles asked, "When could you join a ship?" Woolward didn't hesitate.

"This evening," he replied.

The job Mangles had in mind for Woolward was as a fourth officer on the *Tweed* which was scheduled to depart Southampton in just three days. That didn't leave much time but it was enough for Woolward to buy new brass buttons for his jacket and a gold band to adorn his cap before setting off on the 80-mile journey to Southampton.

Shortly after arriving, Woolward headed for the docks to find the *Tweed*. To the eager 18-year old, the ship seemed extraordinarily large. As he stepped onto the main deck by the starboard gangway he was so disorientated by her size he wasn't sure which way was forward and which was aft.

First contact with the crew would do little to make Woolward any more comfortable. As he commenced with introducing himself, he was taken aback when the midshipman of the watch launched into a string of verbal abuses aimed at him and the company.

It was not the welcome Woolward had expected. As it happened, the midshipman had been hoping to be named fourth officer himself and regarded Woolward as the enemy.

Considering himself more than a match for the competition, young Woolward shrugged off the tirade and moved on, introducing himself to the ship's senior officers. They were, by Woolward's account, impressed by his "hard experience," his "cheek" and the fact that his father had been both a Royal Navy man and the Ramsgate harbormaster. In short order it was settled. The captain of the *Tweed* announced Robert Woolward would be the ship's new fourth officer.

On October 17, 1844, as he prepared to set off for the first time on a Royal Mail ship, Woolward was treated to a grand luncheon served in the *Tweed*'s saloon. There was plentiful liquor and congratulatory speeches by Royal Mail managers who were thrilled with the ship's register—100 passengers—the most yet for the growing company.

Woolward often noted occasions to celebrate, that is, to have a drink or two. Nothing in the West Indies, he would say, was "put through satisfactorily without being kept moist."

For the next several years, the warm waters of the West Indies and South America would be Woolward's proving ground. He cheerfully confesses to bending the rules more than once, apparently unburdened by regrets or consequences.

There were, however, some potentially career ending close calls. On one occasion Woolward was on deck watching a flock of long-billed Booby birds soaring above the ship. The colorful birds are famous fishers that hunt by diving down into the sea after their prey. They are also easy targets. Woolward decided to have a little fun by taking a shot at them. He went off to the purser's cabin where he found a couple of pistols and was in the process of loading one when suddenly, the gun fired. Woolward realized instantly that he was unhurt but he wasn't sure about what had happened to the ball. Following its likely trajectory, he found himself on deck where two ladies were seated, one with her head resting on the shoulder of the other. The shot had barely missed them, passing through the bulkhead, right between their finely-hatted heads.

"Of course, they heard the noise," Woolward recalls, "but I was sharp enough to whip the cork out of a large Eau de Cologne bottle, and plug the hole up with it; having done this I bolted...After all was quiet for the night, I got some putty from the carpenter, made the hole good and gave the place a rub of paint."

The quick-thinking Robert Woolward rose swiftly in the company ranks and by the time he was 28 years old he was promoted to captain.

Around this time, in the British colony of Demerara (now Guyana) Robert Woolward met the woman he would marry, Helen Pierce.

Helen was a no-nonsense looking 19-year old woman with an oval face and dark straight hair worn parted in the middle. One imagines if anyone could draw a smile out of Helen it was the amiable Captain Woolward.

Robert was a man with a reputation in Demerara as the "handsome Captain Woolward," although he wasn't good looking in the usual way. He was not particularly tall. His nose and ears were a bit too large. But overall his face, framed by exuberant sideburns that extended just below his chin, was pleasing. And his eyes, deep set with bushy brows, eyes that could be piercing in anger and dazzling in laughter, were the kind of eyes that drew you in.

In 1854 Robert and Helen married. In December of that same year their first child, Elizabeth, was born. Later came Robert, Helen and the youngest, Frederick who, when he was 8-years old, traveled with his father on one of his cross-Atlantic voyages, a trip Woolward swore helped cure the boy of a chronic whooping cough.

Helen and Robert were often separated for long periods of time. Although he rarely gave her cause to be jealous, there was one voyage where the ladies on board caught his eye. They were a group of thirteen young women, all on their way to marry missionaries stationed in the West Indies. What struck Woolward was that none of the brides-to-be had ever met the men they were expected to marry. It was an extraordinary manner of matchmaking, Woolward thought, but efficient.

The first match was made during a short stop in St. Kitts at the unromantic hour of three in the morning. Two missionary men came aboard and Captain Woolward called for the women to line up for inspection.

"I will have this one," said one missionary. The other man then made his choice. As the new couples went ashore to begin their lives together, the remaining women went below, back to bed, to wait their turn. A similar process followed at each island they passed. By the time they left Barbados just two women were left, ladies who Woolward says he would have been quite satisfied with had he been "in the market."

In the course of thirteen years (1852-1865) Woolward would command many Royal Mail steamships including the *Medway, Trent, Solent* and his least favorite, the *Magdalena*. In all those years Woolward never had a major accident at sea.

Then in the autumn of 1865, Woolward was made commander of the newest and finest addition to the fleet—the *Rhone*. She was the prime example of excellence in comfort and performance that the Royal Mail Company was striving for. "We must be so good that nothing can touch us," Captain Mangles told the board at the annual meeting referring specifically to the *Rhone*.

Captain Woolward had high hopes for this magnificent ship with its powerful steam engine and screw propeller, a major advance over the older wooden side-paddle wheels.

And there was something else special about her. Unlike ships of the last 4000 years or so, the *Rhone* was not made of wood. She was made of iron. Great iron plates secured with rivets formed her hull which was divided into watertight compartments.

"The change from the old style of ship to the new," said Woolward, "was much the same as one would experience if having been in the habit of riding a cart-horse you were all the sudden put on top of a racer."

As with other ships, the *Rhone*'s departure was announced in local newspapers:

> *The Royal Mail Company's new screw steamship Rhone, Captain R. Woolward, sailed hence to-day with the Brazil and River Plate mails.*

She carried 92 passengers and a cargo valued at about

50,000 pounds—adjusting for inflation, about 5 million pounds today.

The specifics were listed in *The Times* of London:

> *£8,000 in bar silver, £20,000 in sovereigns, £11,650 in gold coin, English jewelry value £1,094, French jewelry value £6,430 and a full cargo of British and foreign merchandise.*

It was a relatively warm and typically rainy day when the *Rhone* left England for the first time. Captain Woolward steered the *Rhone* out of Southampton harbor past familiar woods and houses, then west through the Solent straights separating the English mainland and the white cliffs of the Isle of Wight and finally out to the toward the widening horizon.

Steaming out to sea, Woolward was at the center of his world, the land rocking slowly away behind him.

The *Rhone*'s first voyage to Brazil and back would take a total of about seven weeks. Brazil was a country familiar to the veteran captain. On past trips for the company Woolward had managed to ingratiate himself to the Brazilian Empress by making it a point to have a large rhubarb tart, a favorite of the Empress, baked on board his ship and sent to the Palace. In return, Her Highness would send him a large cake with the Brazilian arms and colors on it.

On one of the *Rhone*'s next trips instead of pastry, Woolward would be delivering something to delight Brazil's Emperor, Dom Pedro II. It was a surface condenser, a valuable piece of machinery essential for cooling modern steam engines, and it would be Brazil's first.

Upon arriving in Rio, Woolward was surprised to find the Emperor himself waiting to come on board. He was very anxious to inspect the condenser, his new marvelous machine. Down into the *Rhone* 's engine room went the Emperor and Woolward says he felt obliged to follow.

"I don't think I ever felt the heat so much," says Woolward. "His Majesty however did not appear to turn a hair."

Satisfied with the surface condenser, the Emperor departed the *Rhone* with a sincere, "Good-bye, Caballero Captain!" It left Woolward wondering what the Emperor meant by addressing him in such a formal way.

He soon found out. Later that afternoon two military officers returned to the *Rhone* and presented Woolward with a medallion—a six-pointed star encircled by 18 tiny pink roses all topped with a golden crown. It's called the "Order of the Rose" and is considered a high honor in Brazil.

But when the British Minister in Rio learned about it, he was quick to remind Woolward "no Englishman is allowed to wear a foreign order unless it is given for services in the battlefield, and then only by the express permission of Her Majesty the Queen!"

Rules again!

Not surprisingly, the admonition didn't stop Woolward from wearing the Order of the Rose medallion on at least a couple of occasions, prompting one passenger, Lord Charles Bereford, to suggest that to complement the pretty pink colors Woolward might want to switch from his Royal Mail blue trousers to cherry-colored pants instead.

The maiden voyage of the *Rhone* ended on the evening of December 1, 1865 as she returned to Southampton with a cargo of gold, silver, diamonds, cotton and tobacco and 72 grateful passengers, who in the tradition of the time, rewarded their captain with a token of their appreciation.

"We made a much faster passage than usual," says Woolward, "and passengers were so pleased that they presented me with a valuable pair of diamond solitaire earrings for my wife to wear."

The trip may have been a quick one but the *Rhone* had not entirely lived up to Captain Woolward's expectations.

"She was not too successful on her first voyage," Woolward remembers, citing problems with the engine. "We were much troubled with hot bearings and could not run her at the top speed."

And the *Rhone* would need to be operating in top form. Within days of leaving for her second voyage she would face her first true challenge, a deadly winter storm that would test her seaworthiness and the skills of her commander.

Captain Robert Woolward and his wife Helen. Courtesy: Desmond Woolward

2

LAW OF STORMS

January 1866

A few days after Christmas, Captain Woolward began preparing the *Rhone* for her second voyage. On the ship's register were 67 passengers and a cargo of gold coins, silver and jewelry. It was a middling measure of travelers and treasure, however, this trip would be far from routine.

Out at sea, off the northern coast of Spain in the notorious Bay of Biscay, a January gale was gathering strength, a storm which was destined to make the name of one great ship and utterly annihilate another, plunging more than 200 people into the sea and a breathless death.

Captain Woolward was generally not intimidated by foul weather. He had an uncanny sense of storms and a gift for navigating around them. It was so with some captains. The challenge of doing battle with raging seas was part of what drew them to the job in the first place. Judging a storm's course, strength and speed and then making the correct choice about how to use the engine and anchors to either ride out or outrun them, were life and death decisions a captain faced daily. He had to be prepared at all times because in the 1860s there was simply no predicting the weather much beyond the visible horizon—no weather faxes, radios, satellite images or reconnaissance planes.

Captains like Robert Woolward could benefit, however, from at least two important innovative forecasting tools.

One was the barometer. The brainchild of the seventeenth century physicist and mathematician Evangelista Torricelli, this brilliant device was also remarkably simple. At the top of a long glass tube filled with mercury, Torricelli created a vacuum, a space which would vary in size from day to day depending on the atmospheric pressure. Higher than average air pressure meant fair weather. Low pressure indicated storms were likely. By the nineteenth century, barometers in various designs were popular in middle-class homes and used on ships around the world to "tell" the weather.

Along with the barometer sailors had another resource—a book written by someone who could be considered one of the world's first meteorologists. He was a British Army Lieutenant Colonel named William Reid. Later knighted Sir William Reid in recognition of his work organizing London's Great Exhibition of 1851, Reid was the author of a groundbreaking volume with a long-winded title:

An Attempt
to develop
The Law of Storms
by means of facts,
Arranged according to place and time;
and
Hence to point out a cause

for
The Variable Winds,
with the view to
Practical use in Navigation

Often it's simply called— the *Law of Storms*.

Reid was compelled to write the book after witnessing the aftermath of a hurricane which swept over the island of Barbados, flattening government buildings and killing nearly 1500 people.

"I was induced to search every where for accounts of previous storms," said Reid, "in the hope of learning something of their causes and mode of action."

Reid collected information about the path, wind speed and barometric pressure of dozens of storms in the Pacific, West Indies, America, Europe and the Indian Ocean, then painstakingly mapped the data looking for patterns that would help predict the behavior of storms and the best way for captains to steer clear of the worst winds.

Reid noted some general principles. Revolving storms are often preceded by calms and the barometric pressure might actually read higher than normal on the verge of a storm. So-called "gales" behaved much like hurricanes, were rotational (spinning clockwise south of the equator and counter-clockwise north of the equator) and followed similar tracks. In the West Indian waters ploughed daily by Royal Mail steamships, storms generally moved from east to west inclining northerly. And the severest hurricanes would "leave off blowing at the opposite point to which they commence."

Reid went on to set out some of the first practical rules for navigating around a hurricane.

Step one was to gauge from which direction the storm is

coming. One way to do that is to apply something called Buys Ballot's law. Very simply, in the northern hemisphere an observer whose back is to the wind has the center of the storm on his left.

A ship directly in the path of a hurricane would experience steady winds from one direction, then a calm (the eye of the storm), then winds steadily from the opposite direction. If the storm were moving directly east to west, the winds would first come steadily from the north, then after a lull, switch and come steadily from the south. If the storm was drifting north the steady winds would seem to start from the north, northeast. If the storm was drifting south the steady winds would start from the north, northwest.

If the winds were not steady, if they veered, a captain could determine where he was in relation to it. North of the storm, a ship would experience winds coming from the east by north, gradually shifting east by south. A ship south of the storm would experience winds north by west, gradually shifting to south by west.

Reid also described a hurricane as having two sides relative to its direction—the "right-hand" and "left-hand" semicircle. Imagine a circle (the storm) and an arrow through the center of the circle representing the direction the storm is moving. The part of the storm to the right of the arrow is the right-hand semicircle. Today it's also called the "dangerous semicircle" since the strongest wind speed is often found there and because a vessel traveling *with* the wind will be blown directly into the storm's path.

Reid's rule was that a ship in the *right-hand* semicircle should heave-to on the *starboard* tack and a ship in the *left-hand* semicircle should heave-to on the *port* tack to avoid drifting into the eye of the storm and to find the quickest way out.

Knowing which way to turn was critical. Reid cites the example of the *Judith and Esther* which was caught in a hurricane near Antigua. Although the storm passed over the island in just six hours the *Judith and Esther* spent twenty-four terrifying hours inside the storm because the confused captain rode with it rather than steering out of it.

There's every reason to suppose Captain Woolward knew a lot about Reid's *Law of Storms*. He had the ideal teacher—William Reid himself. In an extraordinary stroke of luck, the very first time Robert Woolward encountered a hurricane at sea, William Reid happened to be a passenger on his ship.

Woolward was a third mate at the time, just starting his career with the Royal Mail Steam Packet Company. When it became clear a big storm was brewing he and his fellow officers gathered around Reid eager to learn from him. As the winds grew to hurricane force, says Woolward, Reid "instructed us in his theory of the 'Law of Storms,' and as it was illustrated at the same time, I thoroughly understood it, and have acted upon the knowledge then gained on all needful occasions since, with the greatest success."

There would be many "needful occasions" including one many years later which was recounted in local a newspaper:

> *The wind came whirling, roaring, rolling with as terrible a sea as ever human eye beheld. Captain Woolward calculated from the force and direction of the wind and the sea, according to the theory of hurricanes that the center of the storm lay directly across* (his course) *and he hoped by laying to to escape the worst. The event proved the correctness of his judgment.*

After that particular storm had passed, in the midst of

speeches and toasts, Woolward had told his grateful passengers not to thank him, but instead to thank "the Almighty Power who had in His mercy, revealed sufficient of His laws to enable His creatures to avoid danger."

God had revealed the law of storms to William Reid, and Reid had revealed it to Robert Woolward.

"The credit of this discovery was due to a soldier, and not to a sailor," noted Woolward acknowledging Reid's theory. "It was now happily possible in most instances for the commander of a ship, by the exercise of ordinary foresight, and by careful attention to the indications of the barometer and the course of the wind to avoid the worst fury of a hurricane."

Thanks to his close encounter with William Reid, Robert Woolward would be better prepared for the January gale of 1866.

Besides the *Rhone*, another ship was about to head naively into that winter storm. She was a very fine passenger steamship called the *London*.

Like the *Rhone*, the *London* was a relatively new ship, owned by Messrs. Wigram & Sons which according to *The Times* had a near perfect safety record, having lost only one ship in a century of doing business. The *London* was an iron screw steamer equipped with the standard safety equipment—five compasses, five chronometers, three barometers, 36 rockets and two signal guns. Although the Passenger Act required she carry only six boats, the *London* had seven—two 26-foot iron long boats (finely lined and especially seaworthy), two 26-foot cutters (workhorses often used to lay cable and anchor), two 26-foot life boats and one 24-foot jolly-boat (usually the smallest and used mostly to ferry personnel to and from ship). She also carried the typical and relatively trivial number of life rings or buoys—six.

A few minutes past midnight on Saturday, January 6, the *London* steamed out of southern England's Plymouth Sound headed for Melbourne, Australia. Aboard were 163 passengers roughly divided between first, second and third class. Most of the 89 crew were British but there were some foreign hands from Sweden, Norway, Denmark and Holland. Captain J.B. Martin was from Australia and had commanded the *London* on her two previous voyages.

But this voyage would begin with some very bad luck. In Plymouth, a small boat which had been helping maneuver the *London* inside the breakwater suddenly capsized flinging the pilot and his assistant the into the water. Captain Martin immediately launched two lifeboats but only the pilot's assistant was saved. It was an ominous beginning.

Saturday, day one of the voyage, the weather was fine.

Sunday turned a bit blustery.

By Monday there was a significant change in the weather. The *London* was swaying heavily in rough seas as the wind roared in from the southwest. Waves breaching the deck, spilled down the main hatchway according to second-class passenger James Wilson, "upsetting the tea things."

Tuesday. Day four. After a restless night James Wilson and his fellow passengers awakened onto a storm-tossed day. Everyone was on edge. The crew tried their best to keep the passengers calm, serving meals as scheduled in the ship's saloons.

But the truth of the situation could be found on deck. It was absolute chaos. The *London* was coming apart from the top down. One by one, like great branches of a tree, the foretop mast, top gallant mast and the main royal masts each cracked and broke. The jib boom (a spar that extend out from the bowsprit) also broke and was dangling off the side. The crew worked feverishly to clear the decks.

Captain Martin didn't know how big the storm was or how long it would last but by early Wednesday morning he had had enough. He gave orders to turn the ship around and head back to England. And that may have been precisely the wrong thing to do.

According to an account in the *Dictionary of Disasters at Sea,* the *London* had by this time actually reached the edge of the storm. By changing course, Martin was unwittingly steering his ship back into the worst of the weather.

Wednesday afternoon. Any brief spark of hope aboard the *London* was soon extinguished as the feeble midwinter light retreated and a mournful moaning filled the air. Passengers and crew listened in fear as the wind escalated, howling more furiously than ever. Hatch covers on the cabins, which had been opened briefly to let fresh air in, were slammed tight again as green seas slammed into the port side. Passenger Wilson says everyone on the ship was now absolutely terrified.

Around this time the exhausted crew made a critical mistake. They hauled in the flying jib boom, which had been dangling dangerously alongside, and roped it down near the engine room hatch. But it was not secure. For several hours the boom banged away against the teak wood-framed glass skylight covering the hatch. Then at about 11:00 p.m. the skylight broke open, exposing a huge 12½ x 9½ foot opening near the center of the ship. Instantly seawater began flooding in and within minutes the engine room was under five feet of water. Now there was no way to power the engines or pumps. The *London* was truly in trouble.

Later one crewman would describe seeing the skylight lying perfectly intact on the deck. Others claimed it was broken in pieces and another swore it was nowhere to be seen. In any event, all agree the skylight was not re-secured.

What followed instead were frantic efforts by both the crew and the passengers to plug the hole. Captain Martin called for volunteers. Collect your mattresses and bedding, he told them, and cover over the engine room hatch. They dragged their mattresses up on deck and piled them one by one. On top of the bedding, the crew laid sails, two ladders and a spar and then tied it all down.

They had to keep the ocean out.

They had to try.

Already the water was knee deep in some cabins. About 80 passengers formed a line and started bailing with buckets. They also used hand pumps. But it wasn't enough. The *London* was settling by the stern.

It was one a.m.

Through the small hours of the night the storm raged on, utterly pitiless.

The water inched upward in the ship and as the *London* sank lower. Around dawn pressure on two of the stern ports forced them to blow. More seawater came rushing in, flooding the lower saloon and raising the water level in the engine room to fourteen feet.

Exhausted and somber, Captain Martin stepped into the main saloon where most of the passengers were gathered and made an announcement.

“There is no hope for us I am afraid,” the captain told them. “Nothing but a miracle can save us.”

Some sobbed. Others prayed, calling on the compassion of Christ to deliver them. But there was no screaming according to witnesses. No panic.

The fear of drowning is as common as rain. But it would

be hours still before the *London* actually sank. Hours to imagine what felt like to drown.

The *London* was only vaguely a ship now. She was more like a great granite rock engulfed by the sea, the waves pounding at her.

No use firing the signal rockets or cannon. The weather was too bad for anyone to possibly see or hear them.

The only chance was the boats.

Three were already gone.

The port lifeboat had been blown overboard on Tuesday and two others soon after that. Captain Martin gave the order to try and launch the four that were left.

The storm had not abated. The furious wind and enormous weight of the waves crashing across the deck fought every human effort to unfasten the boats. But at about 9 a.m. the starboard iron boat was finally freed. It was capable of carrying 50 people.

Passengers gathered on deck, anxiously following the movements of five crewmen who took their positions inside the lifeboat and began to slowly lower her.

But the boat fell unevenly. A heavy swell filled her bow, and like a springboard diver, she flipped forward and disappeared below the waves. There was just enough time for the five crewmen to grab hold of ropes alongside and scramble back onto the dubious safety of the *London*.

Passengers who had been intently watching the sailors' efforts were stunned by what they saw. Their one hope had vanished. Slowly they turned and headed below to wait, resigned to death.

There were now three boats left but launching them seemed impossible. Still, Captain Martin ordered the crippled *London* maneuvered into the wind in a way that was dangerous for the ship but maximized the chances of freeing the boats. For an hour, able-bodied seaman Benjamin Sheals and others worked frantically on the port iron boat but the *London* was heaving so much they couldn't clear her over the side. By 11:30 a.m. they gave up.

But three other crewmen—John King, Daniel Smith and William Hart—had better luck with the port cutter, a lighter boat which is easier to launch. They managed to free her and were now waiting for a smooth patch of sea to lower her. They didn't want this boat swamped like the first had been.

Patience duels with panic in the hearts of men who know their next breath may be their last. They watched. Swells surged and ebbed on the heaving sea. Then at what they judged the right moment, they set the little boat down, swiftly unhooked the tackle, pulled out the oars and pointed the bow into the wind. They had done it.

A handful of the exhausted crewmen, some of them apparently drunk, scrambled into the boat. So did the passenger James Wilson whose timing couldn't have been better. With his cabin flooding fast, he'd fled up on the deck just as the sailors were lowering the boat. He took a chance, he says, and jumped in.

Another passenger, a Mrs. Owen, was also about to get into the boat with her child when Captain Martin stopped her. Midshipman Walter Edwards, who was on his very first voyage and actually knew Mrs. Owen (her cousin was his old schoolmaster), overheard the conversation.

Apparently Captain Martin thought the little cutter was unlikely to survive the storm so he persuaded Mrs. Owen not to board the lifeboat. What he told her was this—staying on the *London* would be "a speedy death" and better than a "lingering death" in the small boat.

And then Edwards says he saw Captain Martin, who had

remained stoically composed until that moment, burst into tears.

A crewman in the cutter pleaded with the captain to come with them but Martin refused. Instead he turned to First Engineer John Jones and, as his last act as captain, gave him the compass heading to land.

As the small port cutter pushed off, Jones could hear the passengers huddled in the saloon praying and moaning.

"I heard no loud sounds," says Jones. "The passengers were very much composed...all that met the ear was subdued groaning."

The men in the lifeboat watched as the *London*'s bow tipped skyward. Their little boat dipped into a trough and when they rose again on the crest of a wave, the *London* was gone.

It was, as Captain Martin had foretold, a speedy death for Mrs. Owen and her child and for the 28 other children, the clergy, lawyers, actors, journalists, couples and crewmen who went down with the *London* that day— 233 people in all.

Huddled together in the small storm-tossed lifeboat, the 16 crew and three passengers who had survived had no idea what fate lay ahead. Land was one hundred miles away. They had among them an odd assortment of provisions—turnips, carrots, a bag of ship-biscuits, a little fresh water spoiled with seawater, two bottles of brandy and two of champagne.

But they were fortunate. After drifting for 20 hours they were spotted by an Italian vessel and rescued.

Later the survivors would tell their "painfully interesting" stories to the Board of Trade investigators who concluded that the immediate cause of the loss of the *London* was "the sea getting into the engine-room and extinguishing the fires." Had that not happened, the report stated, the centrifugal pump throwing 4,000 gallons a minute would have been able to clear the ship. Among the recommendations were to strengthen the hatchways and secure them at all cost from "the intrusion of the sea." The late Captain was not faulted for any navigational or judgment errors.

Of course, the *London* had not been alone in the storm. The *Rhone* had left England three days later on the same general course. Families of the crew knew the *Rhone* must have encountered the killer storm as well. But even after survivors of the *London* were rescued there was still no news about the *Rhone*. Was she lost as well?

Then word came that one ship had spotted her. On January 13, two days after the *London* sank, the *Saxon* had come within hailing distance of the *Rhone*. Captain Woolward, either through signals or verbally, attempted to report his damage and the two ships parted. When the *Saxon* docked at Southampton, her officers repeated what they thought they'd learned. Soon rumors spread that the *Rhone* had suffered heavy losses—16 dead. Alarmed relatives of the passengers and crew were left in almost unbearable suspense waiting to find out the names of those who were lost.

As it happened the *Saxon* had it wrong. The *Rhone* had lost equipment—not people. Later when the *Rhone* made port at Lisbon, Captain Woolward could telegraph the true details to Lloyds of London—three boats lost, two others damaged, two horses killed and one crewman with a broken leg. But there were no deaths among the passengers or crew.

In a letter to the company, Woolward confessed he had "never before experienced such a gale." The barometer had dropped more than an inch in twelve hours to a low of 28.34

and the wind had forced him to lie-to (remain nearly stationary) for ten hours.

Newspapers played up the welcome good news about the *Rhone*. The popular and capable Captain Woolward was the hero of the day:

> *The Rhone it appears was exposed to the fury of the terrific gale which proved so disastrous to the ill-fated London and could not have been far distant from that vessel when she became a living tomb of all on board. The greatest confidence was felt by all who knew the command of the Rhone was entrusted to Captain Woolward, one of the best commanders that ever trod a quarter-deck.*

The *London* sank. The *Rhone* didn't. Whether it had to do with Captain Woolward's skill, the *Rhone*'s seaworthiness or just good luck, the contrast in the two stories secured the *Rhone*'s reputation as one of the safest ships on the sea.

3

DANCING ON THE SEA

Mid-ocean, on sky colored seas, the *Rhone* steamed forward bullying her way across the great Atlantic. At times the bow of the ship launched into the air and then came crashing down again. Assuredly she advanced, transforming the waves around her into tame ripples that tumbled and receded in her wake.

It was wonderful.

But for those passengers who found all the rolling and heaving horribly nauseating it was also awful. A sailing ship at least might cruise with the breeze, its sails pinning the ship down in rough seas to steady it. The steamship pounded away—up, down, sideways—headlong into the wind.

Well, what didn't kill you just might cure you, according to Captain Woolward who had full-hearted faith in the restorative powers of sea voyages. He had personally witnessed many a fellow countryman come aboard his ship in a sickly state, then after the voyage depart revived and refreshed after a few weeks of sun and fresh salt air.

Playing off the fact the company's ships were named after rivers, Woolward would declare, "I am always telling folks of the poor man who had to take seven dips in the river Jordan before he could get cured, whereas only one trip in the Royal Mail Company's rivers is required to set them up completely." He wasn't just saying so to promote the company, he insisted. "It is merely good advice and much better than doctors."

The captain had more reason than ever to feel confident. Ever since the headline-making January storm which sank the steamship *London*, the *Rhone* had become regarded as a favored ship. The *Rhone* was lucky.

In the months following that narrow escape in the Bay of Biscay, the *Rhone* sailed four more times to South America. She was functioning reasonably well, burning much less coal per voyage than previous ships on that route but averaging less than 10 knots per hour, still not top speed.

Then in March 1867, following a series of inspections and authorizations, the *Rhone* and Woolward were reassigned, promoted really, to the West Indies route.

Caribbean bound passengers would have felt particularly comfortable on the *Rhone*. Her wide decks were relatively clear of awkward storehouses and equipment and well-suited for ladies and gentlemen to enjoy their daily sweep of the ship. For those not in the mood for a stroll, lounging on one of the wooden slatted deck chairs positioned under canopies to provide shade from the tropical sun, was an inviting option.

Ship passengers have always been a curious breed, especially curious about each other. Almost immediately after leaving port the games would begin.

Who is he? Where are they from?

Those attracting the most intent gazes tended to be men

and women traveling together. People whispered, trying to guess their relationship to each other.

What a handsome gentleman! And the lady with him... his wife? Fiancé? Sister perhaps? She is young. Young enough to be his daughter!

In such cases the ship's stewardess was considered the best source of information.

"If the gentleman is very attentive to the lady then they are going to be married," declared one such stewardess. "If the lady is very attentive to the gentleman, then they have just been married. If they do not seem to care any thing about each other at all, then they have been married some time!"

This floating village aspect to traveling by ship presented a tempting opportunity for some to exaggerate their position in society. Captain Woolward thought it a futile effort. Ships are too small for secrets to be kept for long.

"Passengers on board a steamship appear to be very different folks from what they are when on shore. But why they should be so I have never been able to make out," Woolward would say. "They seem to forget the world nowadays is very small, and that it is very easy to find out who they are and what are their antecedents."

While most steamship passengers were traveling for business, others decided to book a trip to the West Indies for a very different reason—for the fun of it.

The vacation cruise was a relatively new concept being promoted by the Royal Mail Steam Packet Company. Attracting first-class customers was key to its business plan and that meant providing first-class service. A major draw on Victorian steamers, just as it is on today's multi-story mega-ships, was food.

Mutton chops, eggs, bacon, sausages, Irish stew, curry and rice. And that was just the breakfast.

The evening meal was served à la russe, that is, with courses brought to the table one at a time rather than all at once. A taste for the menu can be gotten by browsing through a copy of Soyer's *Cookery Book*. Alexis Soyer was Victorian England's celebrity chef and a copy of his cookbook was required on all Royal Mail ships. Recipes included 37 different soups like Ox Tail, Red Carrot, Hare with Brown Sugar & Port Wine and dozens of main dishes including Stewed Eel and French Ragout of Mutton. There was also an impressive array of ways to prepare delicacies such as tongue, brains, liver and kidneys. Soyer, ever practical, notes when making Lambs Head with Hollandaise, for example, it was important to be aware that lamb's head takes about half the time as sheep's head to cook.

Company regulations called for the *Rhone* to have both an English and French cook on board and mandated "the tastes and comfort of foreign passengers were always to be consulted and indulged without any extra charge whatsoever."

On Caribbean routes, emphasis was given to the French language and cuisine. A light French soup and at least four French dishes were to be prepared each day. On the Brazil route waiters were expected to speak Portuguese, the bill of fare was to be written in English and Portuguese and local dishes (like black beans stewed with pork and a hot pepper sauce called Molho de Pimento) was to be served on every table.

Passengers wishing to have wine with the evening meal were advised to inform their waiter during lunch so it might be properly chilled for them in advance. Port, sherry, claret, and champagne were available in ample quantities with prices posted in the saloon. Last call was at eleven.

The *Rhone* had two main saloons, one in the bow for second-class passengers and one in the more stable stern for first-class. No officers other than the captain, chief officer, chief engineer, purser and surgeon were allowed to mingle there with passengers. Seating was on benches at long tables and arranged according to cabin number. Hanging close at hand on racks above the tables were a large assortment of bottles, tumblers and wine glasses.

Dining at sea clearly appealed to Captain Woolward who describes suppers that would start at seven and could stretch on until two in the morning and where quarrels were settled over a few glasses of wine and conversation.

Passengers on the *Rhone* ate on ironstone china produced by G.L. Ashworth Brothers expressly for the Royal Mail Steam Packet Company. Plates were white and ringed with one blue and one thinner red line. Plates, cups, saucers, soup tureens and platters all bore the same emblem—a lion, crown and unicorn—the crest of the Royal Mail Steam Packet Company.

Second-class passengers had similar dishes with one obvious difference. Stamped in black letters at the center of each plate was this: "**2nd cl.**" For those not in first-class it was a daily reminder, as they spooned up the last bit of mutton stew from their plates, of the class difference aboard ship.

Further down the epicurean food chain was the crew. Their daily rations were recorded in a document called the *Agreement of Foreign Going Ships*.

Dockside they had fresh beef, vegetables and potatoes. But at sea, while passengers were being wined and dined, the crew had to make do with salt rations (pork and beef on alternating days), bread, peas, flour, tea, sugar and one gill (a fourth of a pint) of rum per day.

When the steam was up the *Rhone's* firemen and coal trimmers were compensated for their hard labor with an extra serving of oatmeal and (what they probably appreciated much more) an extra half-gill of rum.

Food rationing was strictly enforced by Captain Woolward who records in his March-May 1867 log book that he disciplined a seaman for stealing pork from the ship's store. He ordered the man turned over to local authorities in Jamaica where he was sentenced to 60 days hard labor. On the start of the same voyage, Woolward also fined 36 men (about a quarter of the *Rhone*'s crew) for reporting late to the ship.

Dedicated to the task of preparing the daily meals was a galley, bakery, ice house and pantry. Ships like the *Rhone* were also stocked like a biblical arc with live animals in pens —sheep, pigs, ducks, turkeys, geese and chickens. According to regulations the killing and dressing of these animals should be done at night or very early in the morning so that the decks could be washed and perfectly cleaned by 8 a.m. when the passengers awoke.

Occasionally the menu would be augmented with exotic wildlife from local waters like grouper or turtle. The author of a shipboard newsletter called *Hibernian News* describes the crew catching and serving a porpoise to relatively good reviews. It had, the author wrote, a head that tasted like turtle, ribs like beef and a tail similar to apple tart.

Basics, like milk, were a luxury at sea. Steamships had to carry their own cows.

One of Captain Woolward's more demanding passengers (on a special charter trip) actually took the unusual step of writing the amount of milk he expected for his family—

four quarts daily—right on his ticket. He was a lawyer who apparently considered this a binding contract. It presented a dilemma for Captain Woolward because on that particular voyage there were more than 100 passengers and only one cow. Clearly there would not be enough milk to go around.

A cow wasn't so easy to come by on short notice, so Woolward decided that at the next landfall he would buy two goats instead. It seemed like a good solution but his picky passenger didn't agree. His children required "pure cow's milk."

"I suggested," recalled Woolward, "the contract merely said pure milk, and did not state cow's milk or any particular milk, and that I should consider the contract was fulfilled if I could only give him four quarts of pigeon's milk, provided it was pure." But Woolward's argument did not prove persuasive and so at the next opportunity he sold the goats and bought two cows. Woolward was, in the end, "not anxious to give Mr. Lawyer a case to amuse him in England at the Royal Mail Company's expense."

Although long sea voyages could be notoriously dull, on the *Rhone* at least, not food nor drink nor entertainment would have been in short supply. If they weren't strolling or reading or smoking on deck, passengers could gather in the saloon to play checkers or a game of cards (except on Sundays when it was forbidden). Gambling on board was against regulations and to be, respectfully, discouraged at all times.

Evenings there might be a dance on the quarter-deck or a "Programe of Music" with the passengers and officers signing up to perform. Verses from favorite songs like *The Widow Malone, The Sea is England's Glory* and E*ngland the Pride of the Ocean* would fill the air. Or someone might sing one of the new religious hymns which were starting to compete with more popular melodies of the day. As William Booth, founder of the Salvation Army put it, "Why should the Devil have all the best tunes?"

Imagine the *Rhone* alone on a dark night at sea, golden points of lamplight sparkling from her portholes like earth-bound stars. And blending with the drum of the screw beating the water and the splash of waves against her hull, the sound of human voices, cries of encore and roars of laughter.

It's an enchanting picture. But, of course, they were not all magical moments and not all ships were as built for comfort as the *Rhone* was.

In letters fired off to Royal Mail Company headquarters and local newspapers, passengers complained about bad food, bed bugs and unhealthy air. Conditions could be stifling, especially in rough weather with the ports locked up tight. The smells below deck were an almost unbearable mixture of animal hides, coffee, cocoa and tropical fruit combined with a sour odor from the oil lamps which were extinguished each night.

One passenger, while complementing the food on one ship as "good in quality, well-cooked and super abundant," complained about the lack of ventilation writing, "I who had a cabin on a lower deck often exclaimed, 'Oh! For half the food and double the air.'"

And there were the rats. They ate their way through mail bags and champagne corks and were such a problem the company issued a regulation ordering every commander to have at least one cat on board before sailing from Southampton. While cats were considered useful animals, dogs were banned from Royal Mail ships unless they belonged to one of the passengers and were properly ticketed.

Even on a ship as comfortable as the *Rhone,* the two-week trip from England to the West Indies was bound to have its ups and downs.

Passengers describe a predictable sequence of emotions. Initially there's sheer joy. As the ship leaves port passengers are cheerful and looking forward to a very pleasant trip. Then the weather turns grey and ominous and so does their mood. Gloom is followed by thoughts of doom and some begin to wonder if they will ever see land again. Finally, when they arrive safe at a tropical harbor under glorious skies and peach-colored clouds, negative thoughts melt away in the delightfully warm breezes and the travelers end up thinking they had a very good voyage after all.

Joy, gloom, doom and lastly relief. That's what passengers on the *Rhone* may have felt as they reached the island of St. Thomas for the first time. For those from places like New York or London the first sight of tropical seas can be mesmerizing. The ocean isn't the soupy grey they're used to. It's not one single color at all. It's a dazzling combination of electric blue, turquoise, bright green and indigo.

And the harbor of St. Thomas was certainly picturesque, from a distance. Shaped by the outlines of an extinct volcano, it was protected by hills on all sides forming a nearly perfect circle. Inside was a dizzying forest of masts and ships of all sizes. Tugboats, sloops, brigs, schooners and steamers from England, France, America, Spain, The Netherlands and Germany could all be found anchored here. With as many as 20 ships arriving daily, busy St. Thomas was arguably the most important port in all of the Caribbean. It was also the principle transfer station of the Royal Mail Company. Working on the hub and spoke system, big transatlantic steamers like the *Rhone* would come and go from St. Thomas with smaller intercolonial ships working four main routes from there:

- Mexico

- Bahamas

- Barbados, Demerara, Trinidad and Tobago

-Haiti, Jamaica and Colon; where passengers and freight could then make connections for the west coast of America and South America and New Zealand and Australia.

Arriving passengers who wanted to go ashore at St. Thomas could climb into boats for the short ride to a small pier in the busy town of Charlotte Amalie. Its main avenue was quite charming and lined with shading coconut palms where local women in colorful dresses sold sweet oranges, bananas and sugar cane.

And there was shopping. St. Thomas was a bustling cosmopolitan trade emporium. Stores offered everything from practical light clothing and Panama hats to more extravagant black silk cloaks and belts with pearls and ribbons just arrived from Paris.

Food shops sold flour, corn meal, prime pork, boxes of hams, bags of beans, soap, cherry cordial, onions, lard, butter, tobacco and black tea. One hundred good cigars cost about one pound.

A cacophony of languages spilled through the streets and onto the pages of the local newspaper, the *Tidende,* where on any given day you might see ads and articles written in Danish, English, French and Spanish. But in a testament to the importance of British companies like the Royal Mail Steam Packet Company, the most common language spoken in this Danish colony was English.

Despite all it had to offer, St. Thomas was not so much a destination as it was an unavoidable pit-stop. While the streets for the most part were clean, the harbor was filthy.

Passengers reported seeing garbage and the bloated carcasses of dead animals floating by their ships. The stench was so unbearable they felt compelled to slam their ports closed even though it meant cutting off the one source of fresh air. Adding to the malodorous mess was human waste. Raw sewage from the town poured directly into the bay and could remain stagnant near shore for months until, as one passenger put it, "Providence sends a tornado or hurricane to clear away the filthy accumulation."

Other islands were much more tolerable. At each stop the first order of business was for the captain to dispatch a boat to deliver the mail. Passengers would then catch another boat to go ashore. Landings could be tricky with boats sometimes rolling in heavy surf before beaching. An hour before departure the captain would fire the cannon to signal it was time to return to the ship. A passenger could arrange to stay overnight but accommodations on the islands were not always easy to find. Sometimes there were inns, but often there were not and travelers had to arrange to stay in people's houses instead.

Ice was surprisingly plentiful, both a luxury and necessity for preserving food. Imported from America on ice-ships, it was packed in blocks carved from New England's frozen lakes including Walden Pond. In St. Thomas, the Danish government allowed the monopoly to one icehouse which could be fined heavily if it ever ran out.

In order to return to England, passengers would need to make their way back to St. Thomas and transfer to a large transatlantic steamer like the *Rhone*. Thanks in large part to this constant stream of tourists, officials and businessmen and to the money the RMSPC spent on repair shops, coaling berths and wharves, St. Thomas was flourishing. But from a long-term point of view, this busy practical harbor had three major drawbacks.

One was earthquakes. It was not at all unusual for tremors to rattle the little island, shocks which were reported regularly in the local newspaper. Serious damage, however, was rare.

Hurricanes were a much bigger problem. The dangerous storms were so frequent during the summer that churches flew open their doors at the beginning of each season for parishioners to offer special prayers asking God to spare them. But there had not been a truly bad hurricane in the area in 30 years.

St. Thomas was problematic for yet another reason—disease. In fact, it was said if a traveler could smell stench in St. Thomas harbor “he had better go below and make out his will.”

Cholera was a major worry but the disease people feared most was yellow fever. In 1867 no one understood what caused yellow fever, how to treat it or how it spread. All they knew was it was a horrible way to die.

Sometimes referred to as “black jack” or “the black vomit,” yellow fever has three stages.

In stage one the symptoms are ambiguous—fever, headache, severe pain in the legs, arms and back and vomiting.

Three or four days later the symptoms seem to disappear. That’s stage two, the so-called "stage of calm." The lucky ones recover.

The unlucky enter stage three and it happens quickly. Within twenty-four hours patients suddenly appear jaundiced, their skin and eyes taking on a yellowish hue. Patients develop a very high fever and crippling stomach pain with nausea. Then comes the other telltale sign of the disease, the dreadful black vomit caused by bleeding in the stomach. Sometimes patients also bleed from the mouth, nose and eyes. They have seizures. The suffering can continue for 10-14 days, after which time about half the patients die.

Hot weather harbors with their filthy polluted water and putrid air (conditions that were so obnoxiously obvious in St. Thomas) were thought to be the source of yellow fever. The disease was also believed to be exceptionally contagious—spread on contact with victim's clothes and bedding. Neither was true.

It would take many years of research and human sacrifice to find the real cause of yellow fever. The first step came in 1881 when Cuban physician Carlos Juan Finlay published what many at the time considered a ridiculous theory—that a tiny common insect, the mosquito, was responsible for transmitting yellow fever. The medical community rejected his idea but years later an American army surgeon, Walter Reed, would prove Finlay right. How he did it is fascinating but also ethically debatable.

Reed knew solving this medical mystery would help save countless of his own countrymen's lives. For decades hundreds of thousands of people had suffered through yellow fever epidemics in American cities from New Orleans to Boston. Controlling the mosquito population might be the way to fight the disease but skeptical authorities wouldn't invest the time or money to do that without proof.

Animals normally used for laboratory experiments, such as rats, were not susceptible to yellow fever. Reed needed to find another way to test his theory. And he did. Using people.

Reed's classic experiment took place in 1901 on the island of Cuba where America had recently fought in the

Spanish-American War and where many more soldiers died from yellow fever than from battle wounds. Reed had two test buildings constructed. One was called the "Infected Clothing Building." For twenty nights, an army doctor and two other dedicated volunteers slept among pillows, sheets and clothing saturated with the blood and vomit of yellow fever victims.

Another building was called the "Infected Mosquito Building" and was separated by a screen into two sections. One side was swarming with mosquitos that had been allowed to feed on patients with yellow fever and the other was mosquito-free. Of all the volunteers, only the one in the infected mosquito section of Building Two got sick. Reed had proved his point. Mosquitos, not contaminated air or clothing caused yellow fever.

Although none of the volunteers in that experiment died, subsequent experiments to develop a kind of vaccine using infected mosquitos to induce "light cases" of the disease proved fatal to three of the volunteers.

Among them was Clara Maass, a 25-year old American nurse from New Jersey.

According to an article in the journal *Military Medicine*, Maass was a dedicated nurse who was eager to become immune so that she could better care for her patients. She was also financially supporting her mother and the fee she would receive as a volunteer ($100) would help supplement her income. Maass was among a group of 23 volunteers who agreed to be exposed to yellow fever in these experiments. One did indeed develop a light case and recovered, but apparently Clara Maass and two others were exposed to a particularly severe strain of the disease. Maass was bitten on August 14th, became ill on the 18th and died on the 24th.

Maass' death sparked a public outcry and several newspapers denounced the human experimentation with sensational headlines:

"Miss Maass Braved Yellow Fever because of the $100 that was needed "

"Sold her Life for $100"

Dr. Walter Reed also concluded the three fatalities were unacceptable and the human experiments in Cuba were terminated. Mosquito eradication programs, however, were stepped up and soon Cuba was virtually yellow fever free for the first time in 150 years. By the 1930s researchers had also developed an effective vaccine.

But even today treating yellow fever is difficult. There is a recommended course of treatment—oral or intravenous fluids and antibiotics—but there is no cure.

In March of 1867, as Captain Woolward approached St. Thomas harbor for the first time on the *Rhone,* he knew it was beginning to look like this was going to be another bad year for yellow fever. Deaths from the disease were on the increase.

So far most of victims had been sailors. In fact, this outbreak seemed different from others in how it zeroed in specifically on crewmen, while locals and officers and passengers seemed unaffected. Not understanding the true cause of the disease, people speculated that officers and passengers were simply more prudent in what they ate and drank and benefited from cooler, better ventilated cabins.

There were calls from passengers and shareholders to stop using St. Thomas altogether because of the yellow fever problem. The RMSPC stood firm but compromised. It would keep St. Thomas as a hub but set up a temporary coaling station about 25 miles away on nearby Peter Island in the British Virgin Islands. If the rate of fever was high in St. Thomas, ships could refuel at Peter Island instead, reducing the time they spent in the infected area. A small steamer could ferry passengers and cargo back and forth, and since it was only a two and a half hour trip and would only run in daytime, the company promised passengers would be safe and not inconvenienced.

That spring of 1867 yellow fever was bad enough for Captain Woolward to take advantage of this refueling alternative. Although it's not clear from the *Rhone's* log book how long they stayed at Peter Island or how well the re-coaling operation went, the record does show that on the *Rhone*'s voyage back to England there were several illnesses and at least one death of a second-class passenger.

Yellow fever continued to plague St. Thomas in the following months along with shocks of earthquakes, which were mild and not considered serious. None of it would disrupt the contractually-on-time mail packets or dull the spirit of the *Rhone*'s captain. Fond of offering advice and delighting in being right, Woolward also didn't shy away from pulling rank on his passengers. He would have an opportunity to do just that during his third and final trip to the West Indies on the *Rhone* in the summer of 1867.

Among the passengers were two bishops from New Zealand, Dr. Selwyn and Mrs. Harper. They had traveled the land route across Panama and boarded the *Rhone* in St. Thomas for the last leg of their journey to England. Along the way, the bishops had made themselves unpopular with their fellow passengers who privately took Captain Woolward aside to complain. They were hoping Captain Woolward could do something to lighten the mood on the *Rhone*.

According to Royal Mail regulations a captain is "to take care that Divine service is duly and reverently performed on board his ship every Sunday. Should circumstances prevent the performance of Divine service, he is to state the reason thereof at length in the Log Book under his own hand." That much was clear. But how long and what kind of service was to be performed? Regulations were not specific.

In Captain Woolward's opinion there was not just one faith practiced on the *Rhone*. She carried dozens of passengers, many of them foreigners who naturally practiced different religions. The bishops would have to keep their place beside, not over them. All he needed was the right opportunity to make his point, preferably sometime before Divine Sunday.

A few days out, on Thursday, Captain Woolward found his moment. A group led by Bishop Harper was on the main deck singing their church music.

What was "all the noise" about, Woolward asked.

"Don't you know," she responded. "We were practicing for Sunday."

Well, Woolward began, that might not be such a good idea. On the *Rhone* Sunday services started promptly at 10:30 a.m. and were over by 11. There would be no time for singing.

Mrs. Harper listened politely but apparently did not consider this the Captain's final answer. Later that afternoon her colleague, Bishop Selwyn, approached Woolward and asked politely, "Is it true, Captain, we sha'n't be able to have our usual choral service on board?"

"Quite true, my lord," Woolward responded. "My experience is against such things on board, and it can't be except in the evening, as the crew have to change their clothes and have their dinners before noon; at all events, as many of them as have to go on watch at that time."

The Bishop began to protest but Woolward interrupted.

"My lord, you forget that I am the bishop of this diocese,"

Woolward told him, "and I don't allow any interference with my prerogative."

The *Rhone* was Captain Woolward's ship. His domain. His rules. We can assume there were no hymns sung on the *Rhone* that Sunday.

On August 26, 1867, the *Rhone* arrived safely in Southampton on schedule. Along with the two now humbled bishops she carried 172 passengers and gold and silver worth almost $2 million, or more than $30 million in today's market.

That very successful ninth voyage would be Captain Woolward's last on the *Rhone*. After spending almost two years as her commander he was being replaced by someone who Woolward knew, but who up until now had not consider a rival. That man was Captain Frederick Woolley.

The careers of Woolley and Woolward had run relatively parallel to each other over the years, although Woolley was slightly senior. For example, when Woolley was captain of the *Conway* in 1849, Woolward was his second officer. On four occasions Woolley had reviewed Woolward calling him an excellent officer, "very steady and very attentive, deserving of every consideration."

During the Naval Review in 1856 marking the end of the Crimean War, Captain Woolward assisted Captain Woolley on a ship following behind Queen Victoria's yacht. The parade of vessels, all decorated with flags and with bands playing on board, was a grand spectacle and Woolward called the event "a good time."

But why suddenly was Woolley replacing Woolward on the *Rhone?*

One possibility is Woolward had done something to antagonize Royal Mail executives. By Woolward's own account the local Royal Mail agent on St. Thomas, Mr. J.B.

Cameron, never liked him. In fact, Woolward, like other Royal Mail captains, had no real job security. He served at the pleasure of the company directors and could be reassigned or fired at the end of any voyage, something Woolward considered an "unsatisfactory kind of agreement."

Another explanation is that this was simply business usual. According to the *Hampshire Independent,* when the *Rhone* and Woolward were promoted from the Brazil to the West Indies route, the paddle steamer *Shannon* along with her commander Captain Woolley were assigned to Brazil in her place. But since according to Royal Mail regulations "vessels on the West India Route are commanded by senior officers, Captain Woolley was transferred to the *Rhone*."

But Woolward suspects that was not the only reason.

He thinks Woolley deliberately pulled rank and took the *Rhone* away from him.

"The captain of the *Shannon* (Woolley) did not like the Brazil line," Woolward writes in his memoir, "and being my senior claimed the *Rhone* and I was appointed to the *Shannon*." Woolward was apparently angry, in his words, "very much put out by this." But Robert Woolward had little choice. He had to comply.

Together Captain Woolward and the *Rhone* had carried hundreds of passengers and a fortune in treasure across the Atlantic.

The *Rhone* had earned her reputation as a splendid and seaworthy vessel. She was the pride of the fleet and a favorite of passengers. For every one of her transoceanic voyages Captain Robert Woolward had been her commander. Now for the next, her 10th voyage, the *Rhone* would have a very different kind of captain.

4

CAPTAIN FREDERICK WOOLLEY

October 2, 1867
Southampton, England

Under a canopy of typically dreary autumn skies, work continued at the busy Southampton docks. The last of the westward bound letters, packages and cargo along with live hens and geese, bottles of brandy, champagne, soda and other provisions for the voyage were being ferried to the now nearly fully loaded *Rhone*.

As the passengers boarded the ship, Chief Officer Dalby Topper stood by to greet them. All baggage, except carpet bags and hat boxes, would have been loaded on the ship the previous day and passengers were strongly urged to label all of their luggage with their name and destination. Each cabin (painted the same bright flat white) had been inspected earlier that morning by the *Rhone*'s fourth officer and the ship's stewards to be sure they were clean, that the beds properly made and that fresh water had been supplied to each. Waiters, saloon servants and the cooks had boarded the ship four days earlier to prepare their stations and clean the silver plate and glass.

Women passengers boarded wearing stylish framed hoop skirts and men sported peg-top trousers cut tight around the waist and narrow at the ankles. And on every head a hat—caps, fine top hats and small pillbox-shaped bonnets, some with sheer veils that drifted across the face. Some of the passengers might have been invalids expecting a sea voyage to the tropics would help renew their health. Others were on business or relocating with their families.

As the bell rang—"all ashore!"—the sound of laughter and goodbyes. Perhaps some told jokes, others chatted nervously while a few stood by in silence, fighting back a vague sense of fear.

As her engine began to roar to life, the *Rhone's* new captain, Frederick Woolley—dressed to impress in a regulation double-breasted blue coat, gold lace chevron cuffs, black leather boots and a gold-banded blue cap—might have taken a moment to review the crew list, scanning for familiar names. About 30 of the men had served with Woolley recently on other Royal Mail ships. But many of the *Rhone*'s crew would have been less familiar and they in turn might have wondered about the character of their new commander.

Described by friends as tall, dark and manly, 50-year old Frederick Woolley was undoubtedly among the company's most experienced officers. That was reassuring.

But if the temperament of the captain sets the tone aboard ship then the *Rhone* was about to experience a sea-change. While Robert Woolward had been good-humored and mischievous, Frederick Woolley seemed to be quite a different type—dutiful and serious.

A sense of the man can be gleaned from a letter written

just three days earlier on September 30, 1867. The letter is from Frederick Woolley to the Reverend Charles Bellairs, a man who had recently been named to replace Woolley's recently deceased brother-in-law as incumbent in his hometown parish.

Rev. Bellairs had been complaining about the apprenticeship program his son was participating in to become a Royal Mail officer. No-nonsense Woolley basically told both father and son to toughen up.

> *My dear sir,*
>
> *....I can quite imagine your son being dissatisfied with the companions he is thrown amongst but it is the only way to learn his profession, and I never knew a well-regulated mind any the worse for it. It is absolutely necessary the young man should know his profession previous to his going on board a Steamer. It requires Sailors to manage them, but Steamers never will teach young men to become sailors. This is why the directors make it imperative for their officers to be a certain time at sea; we formerly had midshipmen, which was luckily done away with. I looked upon it as ruin to a young man, and I am convinced when your son joins this service he will never repent his apprenticeship. I served my time at sea, and no doubt had to rough it more than your son. ...*
>
> *I beg to remain, my dear sir,*
> *Yours truly and respectfully,*
> *F. Woolley*

Woolley made plans to see Reverend Bellairs when he returned from the West Indies, writing he would be back in England in mid-November and in London a few days after that.

It is a meeting that would never take place.

Although at various times a resident of Southampton and Hammersmith, it is fair to say, as it is for most sailors, that Frederick Woolley's home was the sea. Employed by the Royal Mail Steam Packet Company for 25 years, Woolley had crisscrossed the Atlantic dozens of times as captain of at least seven different ships. It was a life in motion that one imagines suited him if for no other reason than as a distraction from an unrelenting series of personal misfortunes he had suffered. By the time Woolley took command of the *Rhone* he was a widower, childless and had outlived most of his immediate family.

These intimate losses, and loneliness perhaps, seem to have infected the middle-aged captain with a stoic fatalism according to longtime friend Spencer T. Hall, who describes his last encounter with Woolley as particularly poignant.

Hall and Woolley had spent an afternoon together sailing on Lake Windermere in England's Lake District. In the flush of twilight as they returned to the water's stony shore, Hall says he was struck by the tender expression on his old friend's face. Frederick Woolley didn't say much, he never did according to Hall, but as Woolley stared at the dimming horizon he seemed to tear-up, wearing what Hall describes as "the earnest look of a man ready to brave any fate in the execution of a duty."

And then Woolley began to speak about loving England, his friends and God. He did so with the tone of a man, says Hall, who "felt he was already upon the horizon between two worlds."

Between two worlds. Is Hall's recounting of this moment influenced by hindsight or did Frederick Woolley, as he stood on the darkening shore of Windermere, have a premonition of the disaster to come?

They had known each other since childhood, Spencer and Frederick, growing up in the small town of Sutton-in-Ashfield in Nottinghamshire.

Their friendship was sealed one day when the boys were playing near Low-Street and got involved in trying to coax a donkey through a white garden gate. Although he would eventually become a respected local writer, as a child Hall lacked self-confidence, considering himself one of the homeliest boys in town. Woolley on the other hand was the take charge-type. Cheerful. Persuasive. Woolley directed and Hall followed—and so did the donkey right through the garden gate.

In Spencer Hall's opinion Frederick Woolley was a natural born leader.

Frederick Woolley came from a family that was large and well-to-do. The proprietors of a local coal mine called Birchwood Colliery, they lived at Sutton Hall, one of the grandest houses in town. Frederick's mother, Mary Ann Woolley, was descended from generations of seafaring men. His father Samuel, described as portly in a well-proportioned kind of way is listed in town records as having the profession of "gentleman." As they strolled down the streets of Sutton-in-Ashfield people would turn out of doors and greet Mr. and Mrs. Woolley hoping for a nod and a smile in return. This grand couple also had, according to Hall, a rather grand family—twenty children in all including Frederick.

"Beautiful in the extreme," were the Woolley girls. The boys were industrious, though not always content, as they pursed a variety of professions. Pale, intelligent Edwin studied to be a surgeon. Alfred worked at a London banking house and had the gift, if you can call it that, of being able to add up three columns of figures at once. And there were the world-travelers—handsome George with his government job in Mauritius off the coast of Africa, and the sailors, John James and Frederick, who both followed tradition on their mother's side by pursuing careers at sea.

Hall recalls merry gatherings with the brothers sharing stories of faraway adventures and the young sisters around the piano singing *Thy Will be Done* and other hymns. But the harmony of the Woolley household was not destined to last. One by one Frederick's siblings died early deaths. Edwin died just before beginning his profession as a doctor. George succumbed to cholera just as he was about to earn a promotion in his government job. And pale Alfred with his pitiful smile, passed away due to a combination of "hard duty, over-wrought sympathy and a softened brain."

By the time Frederick Woolley took command of the *Rhone* he'd lost his father and sixteen of his siblings and was living with his only surviving brother John James in Leamore Villa on Dartmouth Street in Hammersmith, then part of Middlesex County outside London.

Frederick Woolley had had a chance to start his own family once. In the summer of 1854 the 37-year old bachelor married 33-year old Elizabeth Stevens, the sister of the Reverend W.B. Stevens from their parish back home. Woolley was living in London at the time but after the wedding the couple set up house in Southampton where Hall went to visit. Not fancy or luxurious, "temporary lodgings" is the way Hall describes their home.

During the visit Frederick Woolley took pleasure in leading his friend down to the docks to tour the ship he was commanding at the time, the *Atrato*. And she was something to see. The largest paddle wheel steamer in the world when she was launched in 1853, the *Atrato* was a gigantic 350-feet long with four decks and space for more than 200 first class passengers. She was also the first iron ship built by the Royal Mail Steam Packet Company.

The *Atrato* had many admirers. Among them was the Victorian novelist, a prolific author of more than 50 articles and books, Anthony Trollope. In his travelogue, *The West Indies and the Spanish Main*, Trollope describes traveling across the Atlantic in the "good ship" *Atrato*, Frederick Woolley in command. He enjoyed first-class service from stewards, snacking on fruit and nuts and dining on leg of mutton and iced claret. By day Trollope relaxed on deck under an awning which protected him from the tropical sun. At night, he stayed up playing whist until the sergeant at arms extinguished all the lights promptly at 11 p.m.

While indulging in these luxuries, Trollope admits being a difficult passenger, complaining to Captain Woolley about the size of his cabin, slow service and withered apples. Woolley apparently took all the grumbling in stride, conducting himself amiably and professionally. Woolley was "tender to the ladies and cordial to the gentlemen," says Trollope, and by the end of the trip, the chronicler and the captain were "all smiles."

Woolley was very proud of the *Atrato* and relished showing her off to his friend, Spencer Hall. Woolley also presented Hall with a gift, a beautiful cane he had brought in the West Indies. It seems to have been a very pleasant evening. But it was also one of the last nights Frederick Woolley would spend with his wife, Elizabeth. Soon afterward, Hall says, Frederick Woolley was a widower.

Following Elizabeth's death Frederick Woolley was more than ever a company man. And with "looks which grave experience lends the countenance," Hall says Captain Woolley "ploughed the deep...to the satisfaction of all who confided in his good feeling, his consistent love of duty and his skill."

Throughout his career Woolley received good reviews from the company. He was considered "a steady, zealous, efficient officer and an obliging gentleman" who was "liked by passengers."

But while other officers were ranked "excellent," Woolley was described as a "good" navigator. And in an unusual memo at least one Royal Mail executive noted he thought Woolley lacked something in terms of his character, writing, "I should be glad to see a little more Devil in the Captain."

For ten years Captain Woolley sailed a relatively routine series of voyages across the Atlantic to the West Indies on various ships including the paddle wheel steamers *La Plata* and *Shannon*. Now finally in the autumn of 1867—perhaps thanks to some of that "devil" coming out —Woolley was replacing Woolward as captain of the *Rhone*. It would be a fateful and also perhaps last-minute decision because five days before she was to leave for her 10th voyage, at least one newspaper, *The Times*, reported inaccurately that Woolward was still commander of the *Rhone*.

Probably what was important to Woolley at the time was that by transferring to the *Rhone* he would be staying on the best route, in charge of the company's best ship. In fact, it would be his first time commanding a screw, rather than paddle wheel steamer.

Captain Frederick Woolley reported to the *Rhone* on September 26, 1867. By the 30th most of his crew was also on board including two men who had recently served under Woolley—Second Officer W.H. Garrard and 34-year old

Purser John Morrish, a man who would be by Woolley's side until the end.

Most of the crew were veterans of Woolward's *Rhone* including Chief Engineer John Hooper and Chief Officer Dalby Topper.

Woolley had worked with Topper previously and given him a good review but Dalby Topper was not exactly a star. Early on in his career some called him "middling" and "not a smart officer." Still he had improved both his conduct and skills enough so that within five years of joining the company he was promoted to first officer.

Together Topper and Woolley would be responsible for the safe navigation of the *Rhone*. They were to consult each other three times a day (8 a.m., noon and 8 p.m.) and in cases where their opinion differed, the reasons were to be entered in the log. In situations involving potential danger to the ship, however, once the first officer made his opinion respectfully known, his responsibility ended. The final decision was the captain's.

Aside from the captain, the most respected and often best-paid man on the ship was not the chief officer, it was the master of the ship's powerful and sometimes persnickety steam engine—the chief engineer. His experience and skill was critical to making sure the engine ran efficiently, that parts were properly lubricated, bearings didn't overheat and that critical repairs were made before they became necessary. The chief engineer was also required to keep a strict record of the amount and type of coal used each and every day at sea. Coal was not only expensive, but also bulky, taking up valuable space which could otherwise be used to store profitable cargo. The company was always looking for ways to reduce the amount of coal needed on transatlantic journeys in order to save money.

The *Rhone's* chief engineer was a man named John Hooper. He was in charge of the men in the engine room and the 15 coal trimmers and 15 firemen who worked nearby in the boiler room, a dark cathedral-like place lined with a series of black iron furnace doors below and large boilers above.

The trimmer's task was to position himself inside the ship's cavernous coal bunkers, load the fuel into carts to supply to the fireman and at the same time be sure the remaining coal was evenly distributed so not to upset the balance of the ship. On a typical voyage the *Rhone* required tons of coal. Every ounce of it would have to be handled by these men.

The firemen, dressed in their regulation blue Guernsey frocks with the letters R.M.S.P. CO. on the breast, had the job of stoking and tending the fires. Using pokers to slam open and shut the hot furnace doors, they would shovel, rake and poke the coals being sure to maintain and even and efficient burn. Owing to the extreme strain of their work, instead of four hours on and four off like other sailors, firemen had four hours on and eight hours off.

Among the *Rhone*'s firemen were 30-year old David Reese (who would be one of the last to see Chief Engineer Hooper alive) and two young brothers, William and George Keeping. Like much of the *Rhone*'s crew, the Keepings were from Southampton. 24-year old William and 22-year old George lived at 41 Melbourne Street with their parents William and Sarah, an older sister Martha who worked as a seamstress and William's new wife Rosina who was pregnant with their first child. The brothers hadn't always made their living at sea. As teenagers William had been a dairyman and George a stable boy but by 1867 they were working side by side as firemen on the *Rhone*.

Besides George and William Keeping, two other *Rhone* crewmen lived on Melbourne Street—coal trimmer John Lane who was the father of two baby girls, and Captain of the Hold George Holdeman. Of the four neighbors, only one would survive this next voyage.

Among the entire crew there was just one woman, 42-year old stewardess Janet Inglis, whose job was to cater to the ladies on the *Rhone*. It was rare to have women working on merchant ships. According to log books for the years 1866 to 1871, of 4,500 Southampton crew members only 20 were women. Some were widows of men who had died at sea and were offered work in lieu of receiving the often meager pensions. The pay, about 18 shillings a week, was attractive to women who were otherwise unsupported in the poverty-stricken dockland communities.

Other than Captain Woolley, the person who would eventually become the most famous member of the *Rhone* was not a deck officer or experienced seaman. It was a 16-year old ship's boy named John Bailey.

A handsome teenager with a subtle smile, John wore his thick wavy hair combed back from his face revealing a broad brow and bright eyes. Long the hero of romantic fiction, ship's boys like John were at the beck and call of everyone on board and typically described as eager and adventurous. They were also sometimes cruelly treated, whipped or kicked, and expected to be grateful for the education.

The census of 1861 places John's family in Southampton living at 107 Chandos Street. His father, Henry, is listed as a laborer and his mother Sarah, a wife. He had three sisters (Elizabeth, Eliza and Mary) and one older brother named Charles, who had worked as a sail maker but was, according to John, now working on the Royal Mail steamer *Conway*, which in just a few weeks would be rendezvousing with the *Rhone* in the Virgin Islands.

Ten shillings a week were the wages John earned at sea. He'd already made at least two trips on the *Rhone* that year. On the previous voyage, according to Captain Woolward's log book, John had been "sick off duty" for three days in August. The log entry is signed by three men: Robert Woolward, Dalby Topper and the ship's surgeon Dr. Richard White. There were six ship's boys on the *Rhone*. Coincidentally, four of them were named John—John Bailey, John Askell, John Minns and John Buten. Of the four, three would make it off the *Rhone* alive.

The number of passengers booked on the *Rhone* for her 10th voyage is unclear. According to the *Hampshire Advertiser* the *Rhone* took out "a full compliment of passengers and crew" which could mean more than 300 passengers. Another article notes that 35 of them were ultimately bound for Australia.

In the *Hampshire Independent,* just the basics:

> *The Rhone — Capt. F Woolley leaves Southampton today (Wednesday) with mails for the British and foreign West Indies, the Pacific, Mexico, California, Central America, Australia and New Zealand. Letters, etc. can be posted at Southampton up to 12 o'clock at noon.*

The next line in the newspaper reads:

> *The Shannon — Capt. R. Woolward, will leave Southampton on Wednesday next, the 9th inst, with mails for Lisbon, Cape Verde, ...*

The two captains, Woolley and Woolward, were headed off in different directions and different fates.

That day, the famous *Rhone* seems to have slipped out of Southampton on her final voyage in the unsuspecting calm of routine.

5

BLACK JACK ISLAND

October 14, 1867
Charlotte Amalie, St. Thomas

The *Rhone* responded admirably under her new captain reaching St. Thomas in just 12 days, almost two days ahead of schedule. Captain Woolley writes a comforting letter to his widowed mother describing the voyage so far as splendid and saying that he was looking forward to returning home with much pleasure.

Landfall should have been reassuring sight—the hope of green land rising, pink and yellow painted houses with bright red rooftops climbing up along the steep hillsides of the town.

But on this warm mid-October day, as the *Rhone* cruised toward the harbor packed with steamships and sloops from around the world, instead of the usual excited anticipation about going ashore, one imagines an uneasy stillness on deck.

Black Jack was back.

The yellow fever epidemic, which a few weeks earlier

had seemed to be abating, was as bad as ever. The harbor was a death trap. The *Rhone* would not be staying long.

Like most RMSPC captains, Frederick Woolley had had several nightmarish run-ins with yellow fever. The worst was in the 1850s when he was commander of the RMS *Parana*. To reduce exposure to the disease, Woolley had taken the precaution of coaling the *Parana* far out in St. Thomas harbor, but it would make little difference. One day out at sea, with the ship packed with some 300 passengers and crew, the first symptoms started to appear.

Since yellow fever was assumed to be highly contagious, separating the sick from the healthy was a priority. So was avoiding panic. Some captains in this situation were so concerned about keeping passengers calm that they hid the fact there was an outbreak on board and passengers would only learn the truth after they reached England and health officials placed their ship in quarantine. Or they might, days after returning home, come across the name of a fellow traveler in the newspaper under the heading—"Deaths"—along with the solemn words "perished from yellow fever at sea."

But the outbreak on *Parana* was too widespread for Woolley to keep it secret even if he'd wanted to. One after another people fell ill. Captain of the hold, master-at-arms, a fireman, a female passenger—victims came from every corner of the ship. As soon as someone presented with symptoms they were brought on deck to a canvas tent set up as a makeshift hospital. The ship's surgeon and two passengers who had volunteered did their best to make the patients comfortable, including washing them with a tonic called Sir W. Burnett's fluid, a chloride zinc solution considered to be a powerful disinfectant.

There really wasn't much more they could do.

Of the thirteen people who fell ill on the *Parana*, nine

died including Woolley's personal servant and a midshipman who suffered 70 hours with excruciating seizures before finally drawing his last breath.

All were buried at sea. Woolley would record each death in the pages of his ship's log.

"...passenger departed this life of yellow fever"

"Submitted the body of the deceased to the deep."

"Buried ... with Christian ceremonies."

And sometimes next to each name Woolley would list personal effects:

- 2 jackets cloth
- 2 pair trousers cloth
- 2 pairs trousers white
- 8 shirts
- 2 flannel drawers
- 6 pairs of socks
- 12 neck ties
- 2 pocket handkerchief
- 1 bible
- 3 small books
- 1 writing case
- 1 hair brush and comb

This latest yellow fever outbreak in the autumn of 1867 came at a tender time not only for Captain Woolley on his first voyage with the *Rhone* but also for the Royal Mail Steam Packet Company and the island of St. Thomas.

In fact, on October 16, just two days after the *Rhone* arrived at St. Thomas, a scathing article was published in the local newspaper, the *Tidende*, blaming the company for mishandling a recent outbreak aboard Woolley's former favorite ship, the *Atrato*.

The *Atrato* had spent a wet and rainy week in port where more than a dozen of her crew fell ill with yellow fever. But instead of taking the sick off the ship for treatment (which was company and island policy) they had been treated on board as a cost saving measure until it was too late. Many were so weak they died within 24-hours of arriving at the hospital. The chief officer died while still on board.

Then, despite the illnesses suffered by the crew, the *Atrato* was loaded with passengers and sent off to England. On the way home 40 more people fell ill. Ten died.

The *Tidende* predicted the *Atrato's* homecoming would cause yet "another outcry against St. Thomas and that unfounded statements regarding the health of this place will creep into corners of newspapers and cause an unnecessary state of alarm in the public's mind."

It didn't take long for the *Rhone* to suffer her first victim —Dr. Richard White, the ship's somewhat controversial surgeon. The 36-year old doctor had earned mixed reviews from captains with whom he served over the years, including Woolley and Woolward.

Woolley rated Dr. White's conduct "good." Woolward rated him "satisfactory."

Another captain commented Dr. White was "not very successful in his treatment of the sick."

In fact, Dr. White had served on the *Rhone* on her maiden voyage to Brazil in 1865 where an altercation with a passenger led to his being reprimanded and removed from the *Rhone* by order of the Court of Directors.

But by March of 1867, Dr. White had been reassigned to the *Rhone.*

Captain Woolley had a decision to make. He wouldn't want to repeat what happened on the *Atrato*. So far Dr. White was the *Rhone*'s only casualty. It seems Woolley came up with a compromise. Dr. White was taken off the *Rhone* but instead of going to the hospital he was brought to the private home of King's Physician Dr. Magens for treatment.

As for the rest of the crew, Woolley would, as Captain Woolward had done several months earlier, take the *Rhone* to the company's temporary coaling station at nearby Peter Island. As long as the weather was calm, coal and other provisions could be brought to Peter Island on hulks and lighters. The re-coaling plan was remarked on with approval by the editors of the local newspaper.

The bottom line, saving lives saved money.

And the company was profitable. In 1863, it had signed a new contract with the British government guaranteeing it hundreds of thousands of pounds in subsidies. And although the company had lost a total of ten ships to storms, fire and accidents, there had been no major incidents since 1858. Chairman Charles Mangles could afford to put a positive spin on things at a meeting of the Royal Mail's Board of Directors.

"The satisfactory state of the company's insurance fund," he reported, "showed the great care taken by the captains and officers of their ships. The liberal bonuses to shareholders out of that fund resulted chiefly from the selection by the Board of careful commanders, who were requested on every voyage to keep time if possible, but above all thing to take care of the ships and passengers."

Captain Frederick Woolley was one of those "careful commanders" according to Mangles, trusted with the company's star vessel and the lives of all on board. Woolley would keep the *Rhone* at St. Thomas for several days, discharging the outgoing mail and passengers and loading supplies, and then move to Peter Island. There, far from the yellow fever outbreak, the *Rhone* would be refueled and rendezvous with other Royal Mail steamers to take on the bulk of the Europe-bound mail, cargo and passengers.

The fever was not the only thing rattling nerves, however. In August there had been reports of repeated shocks of earthquake felt in St. Thomas. But what did it mean? Some wondered if earthquakes and that other plague of the tropics, hurricanes, might be related.

There would be this kind of reporting linking the two:

> *At St. Thomas there had been heavy rains and sultry weather up to the night of the 24th when a slight shock of earthquake took place, and the weather cleared up.*

What Captain Woolley could not have known was that while trying to keep distance from one threat to the *Rhone*, he and his ship were on a collision course with an even more menacing one.

6

SAFE HAVEN

October 27, 1867
Two days before the great hurricane

Captain Woolley had about two weeks to get the *Rhone* ready for her return voyage to England. Food, ice and wine would need to be brought on board. It was Chief Steward Christopher Story's job to arrange for the company's linen store to exchange the *Rhone*'s soiled linens and bedclothes for clean ones. Other than Dr. White, records indicate there were no additional cases of yellow fever on the *Rhone*.

Although she would pick up most of her Europe-bound passengers at Peter Island, there were some unscheduled passengers who would join the *Rhone* first while she was still in St. Thomas.

Four were invalid sailors from the HMS *Doris*: Isaac German, Charles Peek, Benjamin Hough and ship's surgeon Dr. Henry Arnot. It is unclear from records how ill these men were, however, it's likely that they were grateful to be returning home on a ship as comfortable as the *Rhone*.

Another unscheduled passenger was a man named John Reid who had been injured in a fall on the Royal Mail Steamship *Tyne.* He reports coming on board the *Rhone* October 15. He would be an important witness to the deadly drama about to unfold.

According Lloyd's agent in St. Thomas, the plan was for the *Rhone* to move to Peter Island on the 24th but Reid says she actually left three days later.

On Sunday morning, October 27, the *Rhone* left St. Thomas for the last time. The trip to Peter Island would take about three hours—a pleasant journey in blue-green, island-studded seas.

Escorting the *Rhone,* a chorus of whispering whitecaps. They ripple and crest, rise and die away, as if chanting in some secret language.

For centuries the vast unsounded seas have inspired deep emotions in the extreme—from lofty love poems to scary horror stories.

There's the legend of the Flying Dutchman, a cursed vessel of fever-plagued sailors denied entry to any port and destined to sail forever around the Cape of Good Hope. And there's Samuel Coleridge's ancient mariner who encounters a ghost ship haunted by "Death" and a red-lipped, leprosy-skinned woman called "Life in Death."

As with most legends, phantom ships were only half imagined. Even Captain Woolley once, while homeward bound from St. Thomas, had had his own strange encounter with one. It was a large sailing vessel adrift in the middle of the Atlantic. Every sail was torn to ribbons, the mast was broken, the rigging wilting over the side and her decks abandoned. Woolley sent his chief officer to investigate. But there were no sailors, no passengers, no one at all. And there were no documents or logs explaining what had happened.

All Woolley could determine was the strange ship's name—*Rover's Pride*. The fate of her crew remained unexplained.

As he steamed away from the mystery ship, some among Woolley's crew might have whispered to each other that crossing paths with a vessel like *Rover's Pride* was a bad omen. Sailors have always been a superstitious lot, seeing signs (mostly bad ones) everywhere.

Odd numbers and carrying passengers of noble birth were considered lucky. On the other hand—having a priest or woman on board, seeing dragonflies or certain species of birds or leaving port on a Friday, the first Monday in April (Cain's birthday), the second Monday in August (the anniversary of the destruction of Sodom and Gomorrah) or December 31 (the day Judas hanged himself)—were all bad luck.

So was knowing how to swim. It meant you thought your ship might sink.

What might have struck some as quaint, was backwards-thinking foolishness in the opinion of English journalist James Greenwood. The well-known writer had made a name for himself investigating appalling conditions in London's poorest neighborhoods. He had a flair for the dramatic, sometimes working undercover disguised as a tramp. He also had a particular interest in the lives of seamen. Greenwood thought it embarrassing that sailors from a civilized country would choose to spend good money on bizarre lucky charms like a caul (the thin membrane enveloping the head of some newborns) believing it would magically protect them from drowning and at the same time scoff at the idea of spending a few shillings on a modern-day life saving device like a cork life buoy.

"He knows perfectly well, that should he be cast into the sea hundreds of miles from land, and with only his own limbs to support his body, he must drown," wrote Greenwood. "He knows that storm or accident may bring him to this fatal predicament any day or any hour; and in the face of possible death he shakes his obstinate head and will not avail himself of the means of life-saving his friends hold out to him."

Greenwood didn't blame the sailors entirely. Their attitude, he reasoned, was probably influenced by the attitudes of officers who discouraged the use of life buoys assuming they would make the men "chicken-hearted."

One Royal Mail captain told Greenwood quite bluntly, "I like my fellows to understand that they've got to keep my craft afloat, or sink with her. Men will work like devils to save their own lives, but how much respect would they have for my property if, a few miles off shore we were in danger of foundering, and every fellow could dance over the side with one of these nuisances made fast to him!"

There was another reason life rings were unpopular. Using them could be, by the sailors' logic, worse than drowning. Instruments of torture, some called them because they might keep a man afloat for hours just so he could slowly starve to death bobbing on the surface of the sea.

Then again, sailors suspected the life rings simply wouldn't work. On that issue, even Greenwood had to agree, they had a point.

His own reporting would help expose the problem. In a series of tests Greenwood purchased several belts each stamped "all cork." But when he took them apart he found instead of cork they were made of straw and rushes that, under the modest weight of seven pounds of iron, soaked through and sank in just a few minutes.

The *Rhone* did have personal life buoys on board—a

grand total of seven. It was a tiny number relative to a crew of more than 100 but not unusual for those times. Still, within just a few days, most of those life rings would indeed save lives including that of 16-year old ships boy John Bailey.

Peter Island, the *Rhone*'s destination on that rather ordinary October morning, is part of a tightly grouped cluster of more than 50 islands known today as the British Virgin Islands or BVI. The biggest islands, Tortola and Virgin Gorda, are mountainous, tropically green and ringed with white sand beaches. The smallest are little more than scrub and rock.

To the south of this island ring is the Caribbean Sea.

To the north is the vast Atlantic Ocean.

In the center stretches a magnificent blue-water highway about 20-miles long called the Sir Francis Drake Channel.

Over the centuries, farmers, Indian warriors, slaves, plantation owners, ministers and pirates have called these islands home.

In the earliest days, there were the Arawaks. They were by all accounts a peaceful people described by local historian, Vernon W. Pickering, as tall and of reddish brown complexion who usually lived in homes built of straw near the coastlines.

The Arawaks were eventually driven out by the Carib Indians—a fierce tribe with a reputation for traveling from island to island in cottonwood tree canoes, raiding and killing. The word “cannibal” actually has its origins with the Carib Indians. Christopher Columbus who “discovered” the islands in 1493 on his second voyage to the New World, believed the Caribs were man-eaters. But while they may have used human flesh in war rituals, there is little evidence they ate it as a way to satisfy hunger.

The European newcomers easily overpowered the

aggressive Carib Indians. Many were killed. Some fled. Others were forced by the Spanish into slavery in copper, silver and gold mines around the New World.

It was Columbus who gave this beautiful group of islands their name. He christened them "Las Once Mil Virgines" after the legend of the martyred Saint Ursula and her 11,000 virgin handmaidens. Over time the name was prudently shorted to "The Virgins."

In succeeding decades, the Virgin Islands were claimed by the Spanish, Danes, English and the Dutch who are credited with building the first small fort on Tortola.

By the early 1700s it was Denmark which held St. Thomas, St. John and St. Croix (the Danish West Indies) while England controlled the sparsely populated archipelago that included the big islands of Tortola and Virgin Gorda (the British Virgin Islands).

While strikingly serene looking, these virgins would prove difficult to tame. The British held islands, especially, had no great natural harbor to exploit and little mineral wealth or flat land suitable for farming.

Island vegetation varied from desert cactus in the low-lying areas to a dry rain forest of hanging vines, lacy ferns, giant elephant ears, philodendrons, white cedar and mahogany trees which covered Tortola's dominating 1,800-foot peak, Sage Mountain. To make their living, planters cleared forest and brush, constructing a patchwork of terraced hillsides along the rugged backbone of the island. They grew cotton and sugar in the highlands and ginger in the lowlands.

Tortola had several good sugar cane plantations, expensive operations made profitable only because of the sweat and labor of African slaves. Thousands of slaves.

In fact, by the late 1700s most of Tortola's 9,000 residents were slaves. As the century grew older the relationship between slave and landholder would become increasingly precarious with revolts and riots breaking out regularly around the Caribbean. In the Virgin Islands, tensions reached a tipping point in 1811 when Tortola became the scene of a sensational trial involving a prominent landholder named Arthur Hodge who was accused of murdering a slave in a dispute over a mango. It became known as the "Hodge Incident."

Arthur Hodge had been born in Tortola, educated in England, then following the death of his father, retuned to the islands to run his family's large plantation. Alternately described as a man of elegant manners and sadistically disturbed, Hodge routinely abused his slaves for years without check until he was finally put on trial for killing one of his slaves, a man named Prosper.

The main witness against Hodge was Perreen Georges who is described in court documents as "a free woman of color."

"I was present" she testified, "when he (Prosper) was laid down and flogged for a mango which dropt off a tree, and which Mr. Hodge said he should pay six shillings for; he had not the money and came to borrow it from me, I had no more than three shillings; he said to his master that he had had no more money; his master said he would flog him if he did not bring it; he was laid down and held by four negroes, on his face and belly, and flogged with a cartwhip; he was under the lash better than an hour; he then got up and was carried up to the hill; and his master said he should be flogged again if he did not bring the other three shillings; he was tied to a tree the next day; and the flogging was repeated; he was licked so long that his head fell back, and he could not bawl out any longer; I suppose he was faint; I then went from the window, as I could not bear to see any more of it."

Prosper died thirteen days later. But as horrifying as the story of the torture and murder of Prosper was, it was not the worst testimony heard during the trial. Other witnesses related how three male slaves—Jupiter, Tom Boiler and his brother Simon Boiler—were cart-whipped to death. And two female slaves, Margaret and Else, died after having boiling water poured down their throats.

Evidence was presented that Hodge was inconceivably cruel to the children as well, including an 8-year-old mulatto girl, Bella, said to be his own child. Witnesses testified children on Hodge's estate were subjected to something that sounds similar to waterboarding. Actually, it was even more brutal.

First the children were taken up by their heels, then let down head-first into a tub of water and held under until they passed out. Unconscious, they were lifted out, allowed to recover and then forced underwater again. The punishment didn't end there. One witness said after suffering near drowning, Hodge "ordered them to be taken up and suspended to a tree, by their hands tied together and in this situation cart-whipped for some time at close quarters."

As the trial drew to a close, Hodge addressed the jury in his own defense. He admitted that he had at times mistreated his slaves but he denied ultimate responsibility for the death of Prosper.

"Gentlemen," said Hodge, "as bad as I have been represented or as bad as you may think me, I assure you that I feel support in my afflictions from entertaining a proper sense of religion. As all men are subject to wrong, I cannot but say that that principle is likewise inherent in me. I acknowledge myself guilty in regard of many of my slaves but I call God to witness my innocence in respect of the murder of Prosper. I am sensible the country thirsts for my blood and I am ready to sacrifice it."

On the morning of April 30, 1811, after deliberating less than two hours, the jury found Arthur Hodge guilty. Eight days later Hodge was hanged. He was the first, and perhaps the only, British slave holder in the West Indian ever executed for the murder of one of his slaves. Pickering calls the trial "revolutionary for the times...where a white man was proven guilty for the murder of a black man and sentenced to death."

The British Parliament had officially outlawed the trade in African slaves four years before the "Hodge Incident," but it took nearly a generation (with the passage of the Slavery Abolition Act in 1833) before enslaved Africans were formally freed.

Without the backbreaking work of slave laborers, Caribbean sugar plantations became impractical and unprofitable. Most were abandoned by their owners and quickly disappeared under the green island brush.

Then in the mid-1850s, in Tortola's capital Road Town, resistance to an unpopular cattle tax combined with outrage over a white man injuring a black man to spark a revolt among ex-slaves. Greathouse plantations and cane fields which had not fallen to ruin already were burned to the ground. A further exodus of whites from the islands followed.

By the year 1867, life in the British Virgin Islands was relatively peaceful. The population of about 6,500 residents got by growing root vegetables and cotton, raising cattle and small stock and trading with their neighbors in St. Thomas. Although former slaves now owned their own land, it was so divided up most parcels were not worth much. Commercially speaking Road Town was unimportant. "A very small place," says Captain Robert Woolward in his memoir, "and also a very quaint one."

Most Caribbean islands also have their pirate stories and this area is no exception. For decades sea dogs and privateers stalked the sea lanes of the Sir Francis Drake Channel. Smuggling was a reliable source of income. The flat coral island of Anegada, which is surrounded by treacherous shallow reefs, was especially well-situated for this unseemly activity. Ships regularly ran aground there and were quite efficiently plundered.

But the most interesting pirate story involving the BVI dates back to 1750. It involves a captain named Owen Lloyd, 55 chests of silver and a secret treasure cave.

The story begins off the coast of North Carolina where the galleon *Nuestra Senora de Guadalupe* found herself crippled in a storm on her way back to Spain. The Spanish hired men on a local sloop to temporarily store the treasure but instead, lead by seaman Owen Lloyd, they stole it.

With a fast ship and thirteen men, Lloyd headed south to the Caribbean. Along with a fortune in silver coins, the galleon had carried valuable church plate, tobacco, cochineal, and indigo all of which would be divided among the crew once they reached a safe place. When Lloyd reached the Virgin Islands he chose Norman Island, a dry, rocky island across the Sir Francis Drake Channel from Tortola, to discretely off-load the loot. Lloyd and the boat's owner each got five chests and the crew divvied up the rest. Most of the treasure was buried on Norman Island.

But the island was not as deserted as they hoped. When a local fisherman approached asking questions, Lloyd took off with most of his crew leaving three of his men stranded on Norman Island. One made his way across Drake's

Channel to Tortola where he told officials part of the story—the part about salvaging tobacco and dyes—not about the silver.

Curious, the president of the island council, Mr. Chalwells, decided to see for himself. He sailed to Norman Island and to his amazement found men digging up bags of silver dollars. Soon word of the Norman Island treasure spread and it became a local free-for-all with some island officials taking bags of coins and plate for themselves.

The law arrived in the form of Governor Fleming of the Leeward Islands, who promised a one-third finder's fee to those who cooperated and turned in their share. Few did. According to one local researcher, from an original treasure estimated to be worth almost $200,000, about $180,000 remained unaccounted for.

Some have speculated that this Norman Island treasure story was the inspiration for one of the most famous pirate tales of all time—Robert Louis Stevenson's classic 1883 novel, *Treasure Island.*

It was this same Norman Island, with its fabled treasure coves and eerie open-mouthed caves, which Captain Woolley and the *Rhone* would pass by on the afternoon of October 27, 1867 on the way to Great Harbor.

Great Harbor was Peter Island's largest anchorage. It was a pretty place and a sheltered one. Nine muscular hills embraced the horseshoe-shaped bay like protective arms. A cacophony of xerophytic plants including organ pipe cactus, coconut palms, hibiscus and white cedar which locals used for boat building, garnished the landscape.

On calm days, the sea would lay flat and smooth as glass here. The dazzlingly blue turquoise water was so crystal clear, a sailor could see all the way to the bottom to a wonderland of waving purple sea fans, orange sponges and scores of tropical fish.

In the days of the *Rhone*, turtles, Caribbean monk seals (extinct by the early 1900s) and manatees thrived on the islands. They in turn attracted sharks including voracious varieties like Bull and Tiger sharks.

It is possible the *Rhone* arrived at Great Harbor just in time to witness one of nature's local spectacles. A writer from the *Illustrated Times of London* describes it taking place like clockwork at 2 p.m. each day. Swarms of birds would descend on Great Harbor and hover over the bay. At the same time, as if on cue, small fish would jump out of the water "to escape their aquatic pursers." It was a dance of life and death, the small fish trying to escape the predators of the sea only to find themselves in the clutches of the predators of the air.

For ships, there were natural advantages and disadvantages to Great Harbor. There was good deep water, in places about 15 fathoms or 90 feet. But while some of the bottom was smooth and sandy, most was spotted with large coral ridges and Captain Woolley would have to know just where to anchor to avoid becoming tangled on the coral heads. He also needed to consider how to put the *Rhone* in a good position to take on coal from shore and passengers from other ships due to arrive from St. Thomas.

As the sun set on the *Rhone* that day, Captain Woolley would have had no immediate cause for concern. Under the circumstances Peter Island, with its fresh clean harbor, must have seemed a good place. Healthy. Peaceful. Safe.

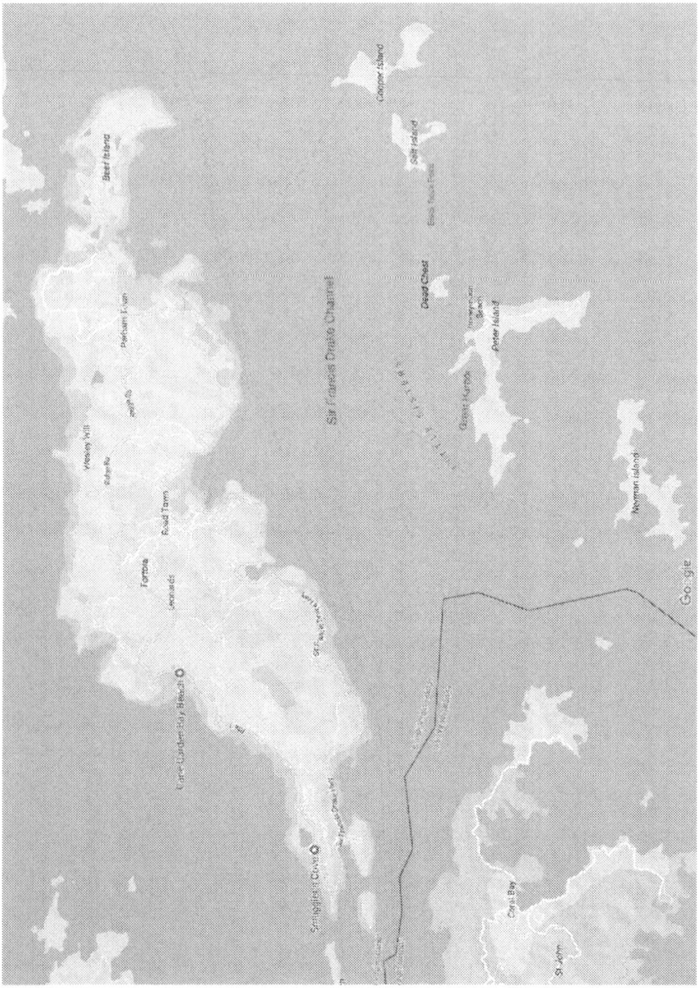
Beef Island
Cooper Island
Salt Island
Dead Chest
Peter Island
Sir Francis Drake Channel
Great Harbour
Norman Island
Road Town
Tortola
Cane Garden Bay Beach
Smugglers Cove
Coral Bay
St John
Google

7

THE HURRICANE APPROACHES

October 29, 1867
Peter Island, BVI
6:00 a.m.

All during the night, from sunset to sunrise, a steady light glowed above Great Harbor. It was the signal lamp of the *Rhone*.

Hanging from the foremast head, the beacon marked the *Rhone's* place in the world. As the sun slowly rose, the *Rhone* herself came into view, blushing in the pastel light of dawn.

The *Rhone* was alone, but not for long.

Gradually maneuvering toward the *Rhone* was her smaller sister ship, the paddlewheel steamer *Conway*. It was the start of what was scheduled to be a busy day. Besides the *Conway* two other Royal Mail steamers, the *Solent* and the *Tyne*, would be heading from St. Thomas to Peter Island for a morning rendezvous with the *Rhone*.

The plan was for all three ships to transfer their Europe-bound passengers and cargo to the *Rhone* which would then head out the following day, October 30, for the two-week trip back to Southampton.

One imagines coming from the saloon of the *Conway,* the comforting sounds of breakfast being served—the clatter of Royal Mail plates and tea cups, the chatter of passengers who may have been, as travelers do, sharing stories of their adventures at sea and plans for home.

There is no official passenger list but according to newspaper accounts there were at least two young couples on the *Conway*—a banker named Arthur John Gibbon and his wife Maria, both in their early thirties and 28-year old James George King and his 23-year old wife Sophia. Also on board were Robert Scott, Henry Arthur Sanderson (a 26-year old officer in charge of her majesty's mails) and 54-year old Benjamin Walrond Esq. of Barbados.

Four other passengers aboard the *Conway* are mentioned but never named. One was an Italian man from Pennsylvania traveling second class. Another was a lady from Demerara referred to only as "Nurse." The woman's husband had recently died of yellow fever and now she and her two children were leaving the heartbreaking tropics behind and heading for England.

The *Rhone's* surgeon, Dr. White, was also aboard the *Conway*. He was not yet fully recovered from his latest bout with yellow fever. The doctor was, as one medical journal put it, "in a weakly and very depressed state of mind," but had apparently insisted on returning to the *Rhone*. Perhaps it was a sense of duty. Perhaps Dr. White was simply eager to leave and the *Rhone* was the fastest way home.

The only passengers on board the *Rhone* at this point seem to be the four invalid sailors from the *Doris* and John Reid from the *Tyne*.

With the two ships lashed together, Captain Woolley

spoke with his counterpart on the *Conway*, Captain John Joseph Hammock. Woolley likely inquired, among other things, about the number of transferring passengers. But there would have been something far more pressing on Captain Woolley's mind on this warm October morning—the weather.

There had been the typical on and off rain showers and the barometer was reading slightly higher than average—30 inches—but the wind was growing gusty and the skies appeared threatening. Great dark clouds were crouching all about and Captain Woolley told Captain Hammock he didn't like the look of it. What happened next, the assumptions Frederick Woolley would make about the weather system approaching and what to do about it, would be critical.

At 8:30 a.m. Captain Woolley ordered chief engineer Hooper to get up steam. Down below in the boiler room fireman David Reese helped light three fires in the furnaces of each boiler, shoveling the coal as evenly as possible to get a steady burn. A short time later, new orders. Light the last boiler fires.

With all the fires now lighted, Reese says, they were reaching fifteen pounds of steam. It would take that much, ten or fifteen pounds of steam, to begin to move the ship. Temperatures in the engine room would have easily topped 100 degrees Fahrenheit, yet it also would have been surprisingly quiet save for the rhythmic hiss of steam and clanking gears.

9:30 A.M.

The *Solent* was now approaching Peter Island carrying passengers and £250,000 worth of cargo for the *Rhone*. This smaller, wood-hulled paddle wheel steamer referred to by one traveler as the "leaky old" *Solent,* was clearly inferior to the *Rhone.* Robert Woolward, who once served as her captain, judged her uncomfortable for both passengers and crew. But in the coming hours this lesser ship and her young captain would prove far more resilient than the *Rhone.*

That the *Solent* could in any way out-do the *Rhone* is something that never would have occurred to Dr. John Wilson, surgeon on the *Solent.* Like others who admired the large handsome steamer and wrote about her in letters and articles, Dr. Wilson considered the *Rhone* "one of the finest ships afloat." Adding to her accomplishments she had, according to Dr. Wilson, just made the shortest passage on record between Southampton and St. Thomas—twelve days, four hours.

But this wouldn't be a day for speed records. By the time the *Solent* arrived at Peter Island the weather was closing in. All the bright colors of the Caribbean were fading to grey and the barometer was slipping slightly to 29.88.

A shout pierced the morning air. It was Captain Woolley. Do not come alongside, he warned the *Solent*'s Captain J.M. Gillies. A gale was coming on.

Gillies guided the *Solent* past the *Rhone* and dropped anchor a distance away. Soon after a boat came with orders from Captain Woolley himself—proceed to Beef Island.

Gillies read the message with concern. Woolley explained he intended to weigh anchor and head with the *Rhone* north about seven miles across the Sir Francis Drake Channel to what he considered a safer harbor at Beef Island —a place with "good anchorage, 16 fathoms of water."

"Look at the chart."

But moving the ships across the Channel was a mistake, Gillies thought. And while apparently the *Conway's* Captain

Hammock had agreed to comply with Woolley's orders and was preparing to move his ship, Gillies decided to ignore Captain Woolley's suggestion and instead remain with the *Solent* at Peter Island. It was a remarkable decision for such a relatively young and inexperienced officer.

Of the three captains—Woolley, Hammock and Gillies —Woolley was clearly the ranking officer. Woolley had been a Royal Mail captain for 19 years, Hammock for 12 and Gillies just five. Actually Gillies had been with the company almost as long as Woolley but it had taken him 18 years to rise from midshipman to captain, earning mixed reviews from his superiors along the way. Woolley's early assessment of Gillies was that he was a "good young officer requiring a little experience." Others described him as quiet, "very sober," and a "fair" seaman but Gillies was considered an "excellent" navigator.

Now the 50-year old Woolley and 39-year old Gillies were commanders facing the same situation but their individual experience had brought them to different conclusions. Gillies thought moving to Beef Island was unsafe. Was it the distance, the available anchorage or something else?

Tiny Beef Island is located just east and very close—shouting distance really—from Tortola. Both islands would have provided shelter from winds from the north, which is where the wind was coming from at this point. Both have harbors deep enough to accommodate a ship as large as the *Rhone*. And although Tortola's Road Town harbor is a bit closer to Peter Island, it's beset by reefs and is relatively narrow compared to the southern coast of Beef Island. Beef Island seemed like a logical choice in a storm.

But it seems it wasn't the quality of the harbors that troubled Gillies, it was the nature of the impending storm.

The weather in the Caribbean had been relatively calm that summer with eight notable storms recorded by NOAA (the US National Oceanic and Atmospheric Association), none of them particularly severe.

Most sailors then considered hurricane season over by October's full moon. Sailors used rhyme to mind the time:

June, too soon;
July, stand by;
August, look out you must;
September, remember;
October, all over

Today the Caribbean hurricane season is not considered over until the end of November. But even with modern day satellite images and computer models, predicting the path and strength of hurricanes is a guessing game. Storms can stall, suddenly speed up or take unexpected turns resulting in the tangled spaghetti graphs broadcast on 24-hour cable stations today.

In the 1860s sailors had to rely on their wits, experience and theories like what William Reid wrote about in the *Law of Storms* to help make sense of the ferocious march of hurricanes across the sea. If this October storm was a hurricane and was moving as hurricanes in this area typically do, generally from east to west, then a ship experiencing a sustained wind from the north would be directly in the path of the storm.

The winds would start blowing steadily from the north and then switch abruptly and come from the south during the second half of the storm.

The *Rhone*'s first captain, Robert Woolward, respected the *Law of Storms* and wrote about it. Her new captain, Frederick Woolley, did not leave behind any such record so we don't know how much he knew or didn't know.

He would, however, have had a copy of the company regulation book with its four-page chapter on hurricanes laying out, though very briefly, similar theories to Reid's. The Royal Mail regulation book also states that hurricanes can occur as late as the first weeks in November and that if a captain encounters one, he should do his best to "get clear of land so as to gain ample sea-room for drifting." If it were not possible to clear land then preparations should be made to use the heaviest anchor with two or three cables attached to ride out the storm.

For some reason, however, on this fateful October morning Captain Woolley did not think this storm was a hurricane.

"It's a norther' brewing," witnesses heard Woolley declare.

A norther' is a sudden storm fueled by cold wind dropping down from the North American mainland. It's the kind of roguish weather seamen in the Virgin Islands would have been very familiar with. If this was a norther,' Woolley's decision to take shelter on the southern side of the northern islands, that is Tortola and Beef Island, would seem to make sense.

But Captain Gillies choose a different strategy. It seemed he suspected this was a hurricane and an imminent one. He decided not to head out to sea or to move to Beef Island but instead to keep the *Solent* near Peter Island. According to the *Solent*'s purser, George W. Knaggs, Gillies dropped anchor about a mile and half from the island—far enough away that she wouldn't be blown ashore and close enough to take advantage of the sheltering hills of Peter Island should the winds shift and start blowing from the south.

At this point on the *Rhone* and *Conway*, there seems to have been no great rush. Luggage was being moved from one ship to another and passengers transferred to the *Rhone* across a gangplank bridging the two ships. Among the passengers were the Gibbons and the Kings, Dr. White and a long-haired sailor with the Royal Naval Reserve named John Metcalf. Metcalf had fallen ill in Trinidad and decided to exercise his "privilege" as a merchant sailor to reserve a space, for the payment of one shilling a day, on any passenger ship he pleased. The ship he chose was the *Rhone*.

According to witnesses and company records, between seven and fifteen passengers transferred from the *Conway* to the *Rhone* that morning. Stepping aboard the *Rhone* at least one relieved passenger was overheard saying to his wife, "Thank God, we are safe here."

One of the last to attempt to make the transfer was the woman referred to as "Nurse" and her children. All of their belongings, luggage and money were already on the *Rhone*. The wind was picking up and she was about to step onto the gangplank spanning the steamers when suddenly the men pulled her back. The weather had gotten too rough. The *Conway* was being forced to cast off. The *Rhone* would be taking no more passengers.

A confusing scene ensued with the two vessels parting so quickly that the *Conway*'s chief steward, Albert Fryer, didn't have time to make it back to his own ship.

It was an unassuming moment which separated a young widow from her possessions and a crewman from his ship but it would mean the difference between life and death. Because she was stranded on the *Conway*, the lady named Nurse and her children would survive the day. And because he was left behind on the *Rhone*, the *Conway*'s chief steward, a husband and father of three, would soon be dead. So would the *Rhone*'s doctor, Richard White. He was returning to the *Rhone* only to go down with her.

Aboard the *Rhone*, ship's boy John Bailey had a brother working on the *Conway*. As the two Royal Mail ships parted, one wonders if John thought he or his brother was the more fortunate.

8

THE EYE

October 29, 1867
10:30 a.m.

The *Conway* was now heading north to Beef Island to weather the storm, but the *Rhone* had not yet weighed anchor. Why? One theory is the *Rhone* did not have enough steam up. But according to fireman Reese, she had 15 pounds of steam which was sufficient to move the ship. Other possibilities are that there was a mechanical problem, more time was needed to raise the anchor or that Woolley was still not in much of a hurry.

But he should have been.

Captain Woolley and the *Rhone* were running out of time.

One after another Tortola and Beef Island to the north, Dead Chest to the east, St. John to the west and Peter Island just a few hundred yards from the *Rhone,* disappeared from sight behind an ominous curtain of rain.

Then suddenly the storm hit, stirring Sir Francis Drake Channel into a hysterical froth of foaming whitecaps.

Exposed in mid-channel the *Conway* was struck full force by the furious wind coming from the north-northwest. From his vantage point on the *Solent,* Dr. Wilson could see the helpless *Conway* fluttering on the water "like a butterfly."

Although not very far apart, the men on the three ships (*Rhone*, *Solent* and *Conway)* lost sight of each other. Even familiar objects on their own ships, like the rigging, faded in and out of view and was replaced instead by the unnerving sound of the wind tearing at them.

The *Rhone*'s port anchor was down with 60 fathoms of cable out to the water's edge and her engines running hard to relieve tension on the chain. To Dr. Wilson on the *Solent,* who at times caught a glimpse of the *Rhone,* it seemed as if she was doing well riding out the storm.

She wasn't.

Captain Woolley feared the *Rhone*'s anchor was slipping and that she was in danger of running aground.

"(We) cannot steam against such a breeze," he was heard calling out.

As the wind continued howling, the barometer suddenly plunged to a worrying low 27.95—a sure sign of a hurricane. Despite this, Woolley still apparently intended to stick to his plan to move the *Rhone* to Beef Island.

What happened next is critical. Eyewitness accounts vary but the story with the most detail comes from someone who should know—the *Rhone*'s 31-year old boatswain Emmanuel Sherwood. Usually one of the most experienced sailors on the ship, the boatswain is in charge of inspecting the ships sails and rigging each morning. He is also responsible for all deck activities such as dropping anchor and handling the sails. Sherwood had previously served on the *Shannon* with Woolley.

With the storm raging around them, Sherwood and Chief Officer Dalby Topper began arguing about what to do.

Weigh anchor, were the captain's orders and Topper was duty bound to carry them out. "Sherwood, I want to slip the leads, we must get our anchor to go over under the weather land."

But Sherwood resisted. "We cannot heave in; there is too much strain upon the cable," he told the chief officer.

Topper was not convinced. "We must hang it to the fore-mast compressor and give cable enough to unshackle and slip."

"No sir, I beg your pardon." Sherwood countered. "I think you had better loose it to the fore-mast bits and ease the fore-mast compressor."

"Very well. Do what you can." But the chief officer had hardly finished speaking when the anchor chain broke.

We've lost the anchor, Sherwood cried out.

"Never!" Topper exclaimed and immediately went off in the direction of Captain Woolley to tell him.

The *Rhone* would have had a second anchor but according to John Metcalf (one of the invalid sailors unlucky enough to find himself on the *Rhone* that morning) the wind was so fierce at this point the crew couldn't manage dropping the other anchor. It's also possible there wasn't enough spare chain on board or that Woolley just panicked.

Without an anchor, staying at Peter Island or heading for Beef Island were no longer options for the *Rhone*. Land had become the enemy. Woolley needed to get his ship as far away from the rocky islands and reefs as quickly as possible.

Captain Woolley ordered the *Rhone* to head out for open sea. Full speed.

Below deck, blind to the weather, fireman David Reese still wasn't worried. He was used to a bumpy ride.

"I've often been out in roughish weather," Reese recalled. "I hadn't been on deck in the wind but below this did not appear worse than I've been out in before...We were told to get up as much steam as we could. We got up twenty pounds, all that we could possibly give her."

But what course to take?

Although the Sir Francis Drake Channel is wide and deep, the islands surrounding it are relatively close together. Between them is a maze of reefs and rocks. South of Peter Island is the wide-open Caribbean and no land for miles but a sailor has to choose the way out with care.

One possible route was to go east about four miles past tiny Dead Chest Island and a treacherous just under the waves obstacle called Blonde Rock, then turn south through the passage between Dead Chest and Salt Island. In good weather, it would have taken the *Rhone* about a half-hour. A quicker route was to go east and then quickly south between Peter Island and Dead Chest Island, through a navigable but much narrower channel.

Most modern-day accounts assume the *Rhone* took the longer route between Dead Chest and Salt Island. But according to three separate eyewitnesses (Dr. Wilson, purser George Knaggs on the *Solent* and Emmanuel Sherwood on the *Rhone*) Woolley actually took the *Rhone* through the nearer channel between Dead Chest and Peter Island.

The wind was roaring as Woolley shouted out orders, struggling to keep the *Rhone* on course. It's likely he couldn't see even a ship's length ahead and reports tell of so much electricity in the air that the compasses were useless, their hands swinging wildly in the storm.

The deck was treacherous with gear flying everywhere. At one point the wind grabbed a lifeboat and sent it like a rocket speeding across the deck striking Captain Woolley full in the back. He was stunned but okay and stayed on deck as the *Rhone*, wind to her port side, continued to steam ahead.

The main obstacle the *Rhone* had to clear was Peter Island. Woolley ordered Sherwood aft to take soundings but when he tried to drop a line the wind simply lifted it straight out horizontally like the rope in a game of tug-of-war.

"What water have you got?" shouted Woolley.

"None at all," Sherwood replied.

Then, even more suddenly than it had begun, the storm subsided. The wind dropped to a breeze, the rain relaxed and the fog lifted. On deck, John Reid and Emmanuel Sherwood could now see what was invisible to them a moment before.

"There is land on the starboard bow," Sherwood shouted out. The shore of Peter Island.

Immediately Woolley gave the order—"Full speed astern!" The *Rhone* responded backing up, says Reid, about 600 feet, twice the length of the ship. Then she began steaming ahead again keeping a safe distance from the rocks.

The *Rhone* had passed Peter Island, narrowly avoiding disaster. She was cruising now at full speed under a clearing sky, battered and damaged but sound. With sun shining through thin layers of clouds overhead, she glided through the narrow, becalmed channel toward the open Caribbean Sea.

12:15 P.M.

It was eerie how abruptly the wind subsided. Perfectly calm, is how Dr. Wilson described this moment. The *Solent*, with two cables down, had survived with little damage. Looking out toward the horizon, Dr. Wilson saw flashes of lightening and heard the distant rumble of thunder. By now it was clear this was not the norther' Captain Woolley had predicted. It was a hurricane. In today's terms it was a Category 3 hurricane with wind speeds of more than 125 miles an hour.

And they were in the worst place imaginable—inside the eye.

A mature hurricane is the most powerful natural event on earth. It can be deceptively small or monstrously large—more than 1000 miles across. Winds range from 74 miles per hour to over 150 miles per hour.

The heart of the storm, is called the eye. It is a calm column of air averaging about 15 miles in diameter circled by a solid wall of intense and super-powerful thunderstorms.

Being inside the eye can be surreal—skies blue but hazy, the light voices of birds floating in the air. "Spectacular" is the word Nathaniel Bowditch, author of the classic *American Practical Navigator* uses.

"The eyes' abrupt existence in the midst of opaque rain squalls and hurricane winds," he writes, "the intermittent burst of blue sky and sunlight through light clouds in the core of the cyclone and the galleried walls of cumulus and cumulonimbus clouds are unforgettable."

The strongest winds are in the inner band of the eye where they chimney upward around the comparatively calm core. Because of how the storm rotates, the winds at the opposite wall come from the reverse direction.

Every hurricane is unique and on its own unpredictable and destructive path. This hurricane of October 1867 was a relatively small and fast-moving powerhouse.

Even now, not everyone on the *Rhone* appreciated the danger. The ship's storekeeper, Charles Berry, was actually relieved and chatting with the ship's chief engineer, John Hooper. As far as Berry could tell they were in the clear. Far to his left he could see the distinctive cluster of rocks off Salt Island called Black Rock Point. At its tip was a menacing looking boulder so black it looked as if it had been scorched by lightning. Hiding just under the surface, a tumble of large rocks and coral cascaded 80 feet down to a sandy bottom. Berry and Hooper were watching the point as they heard Captain Woolley order—"turn ahead, full speed." The *Rhone* continued moving generally south out to sea.

But by this time Woolley must have realized this was a hurricane, they were trapped inside the eye and that the winds were about to shift. If he had managed during that first part of the storm to stay at anchored at Peter Island the *Rhone* surely would have been in a safer position, protected by the hills of the island. Instead the *Rhone* was exposed and vulnerable. She was steaming south, away from the rocky islands but toward the oncoming wall of wind.

On the *Conway,* Captain Hammock was taking advantage of the lull, preparing his ship for the worst and pointing the bow to face second part of the storm.

On the *Solent*, the storm had kept the crew so busy and the visibility had been so bad no one had noticed the *Rhone* was gone. Now off in the distance Dr. Wilson saw her. He pointed her out to Captain Gillies who was shocked to see the *Rhone* on the move, "astonished they should have taken such a step."

The eye wall was now approaching the three ships. As thunder clouds once again began to shroud the island coastlines, the *Solent* lost sight of the *Rhone* in the distance.

The lull was over.

12:40 P.M.

During the lull, which lasted about 25 minutes, the *Rhone* had been steaming at full speed. Assuming she was traveling at about 12 knots (technically her top speed was 14 knots but she had been averaging ten), the *Rhone* would have covered a distance of about six miles during the eye of the storm. She would have made it through the Salt Island Passage, putting her roughly south of Salt Island itself—still very close to land.

Now, with supernatural speed, day became night and a dreadful darkness descended on the *Rhone*.

The wind rewound, howling like a mad-cat. It was coming from the south-southeast, the opposite direction from the first part of the storm—just as the *Law of Storms* predicted. Mountainous waves rose up and crashed over the deck of the *Rhone*. Rain poured down and the air became a suffocating mixture of rain, seawater and hail.

On the *Solent,* Captain Gillies and his officers continued their efforts to secure their ship but they could barely see three feet in front of them and were forced to crawl across the deck on their hands and knees. They had two cables down and the engine running to relieve the strain on the anchor. But these winds seemed even stronger than the first. One by one the masts on the *Solent* began to crack and fall. The wind flipped a mail boat up into the air and off the ship.

On the deck of the *Rhone,* Captain Woolley, wearing a

blue flannel coat with a life belt under it, was struggling to bring the ship "up to wind." At Woolley's side, dressed in his regulation single-breasted blue coat and no life preserver, was the purser John Morrish.

Below decks most of the crew, including John Bailey, were holding on for dear life as the ship pitched wildly. In the boiler and engine rooms the men were still feverishly at work. Reese and his fellow firemen, including the brothers George and William Keeping, were ordered to continue stoking the furnaces to keep the steam up.

About this time, a horrified crew member watched as a heavy wooden pole supporting the rigging came crashing down onto the deck, crushing Darby Topper. The chief officer was killed instantly. He was the *Rhone*'s first casualty.

Seamen George Wallace and John Eaton took the wheel for a time and for more than a half hour the *Rhone* steamed hard on, but she was barely moving. Even worse, she was losing ground. This time her greatest threat came from her port or left side—Black Rock Point at the western tip of Salt Island. They were slowly moving closer to it. The crew on deck might have sensed it, struggled to see it.

Captain of the hold George Holdeman shouted to Captain Woolley that land was close —"Very close...nearly abreast of us midships on the port hand."

Passenger John Reid saw it too. For a few terrifying minutes the rocky point loomed unnaturally near, with every swell of sea, ever closer.

Then, above the shriek of the wind, a dull ominous sound.

Bu-bump.

From down below fireman Reese says he could feel the bottom of the ship scourge against rock on the port side then roll and hit again lifting the ship up.

Bu-bump.

Captain Woolley couldn't believe it.

"Good lord, is it ever possible?"

'Yes sir," Holdeman replied. "The ship's ashore."

The *Rhone* had struck the rocks broadside near the wheelhouse and funnel.

Again.

A deadening thud as she hit the rocks a third time but this time the *Rhone*'s iron hull gave way—caving in. Her engine had been running full speed. Her furnaces, pipes and boilers were super-hot. As cool seawater flooded the interior, some witnesses say they heard the sound of an explosion—the boilers bursting downwards.

Fireman Reese says the rocks appeared to cut up right through the hull and into the boilers. Sea water was rushing in. But he doesn't mention an explosion at the moment of the collision.

What is certain is that the impact of the *Rhone* hitting the rocks of Black Rock Point was felt by every person on the ship—the passengers, officers, stewards, firemen, barman and engineers—more than 140 people in all.

People scrambled up stairways only to emerge into the full force of the hurricane. Looking for a way out, they knocked around the deck as if they had lost all sense of reason. There was no evacuation plan, no general order, no time at all.

"Now my lads you must look out for yourselves, one seaman heard the second officer exclaim as the waves hammered at the broken ship.

John Reid rushed up on deck.

"Mr. Morrish," Reid shouted to the purser, "our only chance is to try and lower one of the boats."

Fighting the wind, they reached one of the starboard

lifeboats, grabbed the ropes and prepared to lower her. But she wouldn't budge. Reid leapt into the boat.

"Jump in and our weight will clear the fall."

It may have been a good idea but Captain Woolley had another. The stern of the *Rhone* was now on the rocks. She was breaking up, in her death throes. It was terrifying and land was temptingly close. In the panic, it seems Captain Woolley was focused on saving himself.

"Come this way, Morrish," Reid heard the captain call out. "I think we can jump on to the rocks."

If this is true, it's a controversial decision—for the captain to abandon ship with passengers and crew still on board. This is what Reid says he heard. But as it happened, Woolley and Morrish never managed to get close enough to try for the rocks. As the purser was running toward the Captain, says Reid, the ship began splitting in two. All Morrish and Woolley could do was cling to the wheel to keep from being swept off. A nightmarish chasm had formed in the middle of the ship and the *Rhone* was swallowing her own.

By now Reid had given up on the lifeboat. Instead he looked around, grabbed a quarterdeck seat and knocked the gratings out. Balancing on a railing, he awkwardly stepped his legs through. It was an unlikely life preserver but it would have to do. John Reid couldn't swim.

A huge breaker, large as a house, was now looming over the ship. Reid tried pushing himself off the deck in his makeshift life preserver but got tangled and he braced himself to be swamped.

The great white wave engulfed the deck. Clinging to the ship's wheel, Woolley and Moorish were struck sideways and washed off their feet over the top of a skylight. With the next wave both men were swept off into the seething white water between the deck and the rocks. George Holdeman says he saw it happen, the very moment Captain Woolley was washed overboard. For John Reid, it's as if the captain had simply disappeared. The wheel was gone. The purser was gone and Captain Woolley was gone.

Reid grabbed the railing as another wave passed over him and then he jumped.

Below deck, chief engineer Hooper called out to his men. There was only one chance to survive. Get out. Get out!

With seawater flooding in, fireman David Reese grabbed the engine room ladder and started to climb. In front of him was boilermaker Samuel Nicol. Behind him were his fellow firemen. Then a steam pipe exploded. Everything was enveloped in a hissing cloud of acid-hot, blinding steam. Reese heard screams from men trapped below. Most couldn't make it out. Two firemen who eventually did, John Lane and Arthur Cull, emerged on deck fearfully scalded and later died.

Reese held the ladder tightly. He couldn't see. Too much steam. But he kept climbing.

"As I knew the way out as well in the dark as in the light, I got hold of the handrail and crawled under the steam out of the door," Reese says. He emerged to another nightmare. "When I got out I found the forward companion covered with people. The main deck people were clinging to everything they saw." Among those holding on for dear life was a young couple.

"I saw one lady clinging to her husband by the after companion," Reese says. They would have been either James and Sophia King or Arthur and Maria Gibbon all of whom had only a few hours earlier transferred from the *Conway* to the *Rhone*.

No surviving crew member describes any effort to save the dozen or so passengers. A British official, Captain Vessey, later reports that John Metcalf told him that passengers were tied down during the storm to protect them from being washed overboard but he is the only one to mention such a thing and it is more likely that the passengers suffered through the storm huddled in the ship's saloon until the *Rhone* crashed and they were forced topside.

It seems during these final critical moments on the *Rhone* there were two choices, if you can call them choices: Hold on to some piece of the sinking ship and hope it floats, or jump.

Most people were simply being washed overboard.

There were only seven life buoys on board and, according to John Reid, Captain Woolley was wearing a life belt under his coat when he was swept overboard. Odds were not good to get one.

John Minns, one of the ships boys, says he begged a fellow crewman to help him. "Look out for me, and get me a life buoy when I jump over."

"Oh my son," the crewman told him, "I shall have enough to do to look out for myself."

It was that way all over the ship. Every man, woman and child for themselves. Most never had a chance, like the fore bedroom steward, William Stroud and his 13-year son William Jr., a dispensary boy on the *Rhone.*

Stewardess Janet Inglis asked ship's boy, John Askell, what she should do. "Never mind: it's all over," he told her and she fainted right away. As for the teenager John Minns, with or without a life buoy, he was determined to do something so he jumped.

Ship's boy John Bailey, who also had been below when the *Rhone* struck the rocks, came running up on deck when the ship hit. No shoes. No cap. Reaching the outside, he was instantly swept off his feet and pinned against the rails. All around him was flying debris and hail as big as marbles. People were shouting and shrieking and the surf was "banging against the rocks and throwing up a great sheet all white as snow and as high as a house or higher."

In the chaos John could make out familiar faces but none, he thought, who could be counted on as a friend. Then he recognized someone he knew very well, a seaman named James Gough who was just four years older than John. Gough was hanging on with one hand, throwing one of the ship's life preservers over his shoulders with the other. In his fright, John blurted out an obvious and at the same time ridiculous question—will you give me your life buoy?

"I didn't expect it, of course," John says.

The storm was so loud John could hardly hear himself speak. Gough either didn't hear John or chose to ignore him because John says Gough "took no notice." Quickly, John had to think of something else. Fighting his way up to the stern John, as Captain Woolley apparently had, decided to try to jump for it. He realized he could easily break a leg or sprain an ankle, but that would be better than drowning.

John braced himself and jumped. And missed.

"I can't tell you how or why I missed but I fell half into the sea and the surf came in a minute and licked me off and carried me out."

Fireman David Reese was half running, and half swimming over the shattered deck. He could see the chief engineer John Hooper clinging to a skylight and was washed by him.

Their eyes met. "He looked at me and I looked at him," says Reese, "just as men will do at such times."

Then Reese himself was washed overboard.

In the next and final moments the middle of the ship sank lower and the deck of the bow disappeared below the waves.

The *Rhone* had hit the rocks, split in two and sank within a time span of about ten minutes.

9

DRIFTING MEN

In the water now, John Bailey grabbed hold of a piece of wood—a "stick" he calls it—which was just large enough to keep him afloat. He held on, jealously guarding it from the thieving waves which seemed at every moment to want to snatch it away. The boy was being dragged forward and back, sometimes very close to the big boulders on the tip of Black Rock Point. He imagined he could reach the rocks and pull himself up but each time he tried he was sucked back again into the churning sea.

The sea was heaving with debris—sharp metal, heavy wood and tangled rigging. Dozens of people were flailing about, struggling to grab anything that could keep their heads above water. Some could swim. Many could not. To Reese it seemed like more were killed by "the knocking about of the wreck than being drowned." Reese, who could swim, managed to catch hold of a sun-sail boom. But then his leg got caught in some rope and he was pulled under.

John Reid was in a similar predicament. His pants were snagged on something, tugging at him, pulling him down. He was drowning. Underwater he tore at his trousers, managed to rip them away at the knee and free himself. Bursting back to the surface, Reid took a grateful gulp of air and saw the wreckage all around him shifting and churning. The forepart of the ship had now drifted back up, but not for long. As it twisted in its sad dance, it began sinking again and Reid was dragged down with it. Again, he fought his way back to the surface. Then he saw the ship's fore mast was sticking out of the water almost upright. Instinctively Reid grabbed for it, dragging himself up.

He was not alone. David Reese had managed to crawl onto the mast too. So had John Metcalf, Edward Barber, and 16-year-old John Askell who had attempted to stuff a handful of oranges and pineapples into his flannel jacket before a wave came and swept him off the deck. The five men, washed nearly naked by the storm, held tight. Reid could see about a dozen others nearby clinging to the stern. One by one they dropped away leaving a single fireman, a man who Reid recognized as one of the Keeping brothers. "He hung on until the sea became calmer, when he seemed to be preparing to swim ashore, as he was pulling off his clothes, then a sea broke over him and swept him away." It was the last anyone saw of him. George and William Keeping, both in their early twenties, William with a baby on the way, drowned that day.

The mast was just a few yards from Salt Island. It was an easy swim in decent weather but in the driving storm the men were too terrified to try for it. And although they couldn't be sure how long the mast would stay upright, they stayed put.

Still in the water, Charles Berry saw what seemed like 30 or 40 people all around him desperately clinging to whatever might save their lives. His life depended now on two small pieces of wood. Only minutes earlier he had been chatting with the engineer, so confident the *Rhone* had beat the storm.

Berry spotted one of the ship's boats which was turned upside down but still afloat. Auguste Gaurion, the *Rhone*'s young fore bedroom steward, was on top of it. Berry swam for the boat, reached it and hung on. Two others followed. The seas were still high but the four men manage to keep hold of the upturned boat.

Meanwhile, ship's boy John Minns, who had jumped before the *Rhone* went under, was straining like the others to keep his head above the suffocating sea. Unexpectedly he found himself fighting alongside two officers for a handhold on an awning boom. Rank had little weight in this situation and neither young John Minns nor the officers could manage to the hold prize. Instead, Minns was swept over to a grating and the older men towards a floating bale of hay.

Minns was close enough to hear their voices. "Give up old boy," the fourth officer said to the second. "I can see it won't do for us."

They let go and John Minns lost sight of both men in the waves.

Those who hadn't drowned or been crushed in the debris began drifting away from the wreck and from each other. John Bailey estimated he'd been in the water about a half hour when he spotted another man in the distance. He recognized him immediately as his friend James Gough, the seaman who had been lucky enough to find a life buoy. Clutching his flimsy piece of wood, John swam toward Gough hoping, he says, for at the least a little company. But as John got closer he witnessed a strange sight. Gough suddenly threw his arms up over his head, slipped down out of the buoy and disappeared.

What had happened? John is not sure but suspects a

shark got him. "Nipped off at the legs, that's how I recon it up."

But John didn't panic. Instead he reached out for the empty life ring. It was large, wrapped in canvas and had the name *RHONE* painted on it in block letters. John managed to lift it over his head. He ran his arms through and slipped into it. His immediate sensation was not relief, however. It was fear. To him it seemed the still raging ocean had an even firmer grip on him now and was carrying him out to sea. John drifted further and further from where the *Rhone* went down until the sound of the waves breaking on the rocks grew fainter and fainter.

The storm, so astonishingly fierce, was also surprisingly short. By about 4:00 p.m. the worst wind was over, though a steady rain continued. John Reid and four others perched on the *Rhone*'s mast had climbed as high as they could but still they were only about four feet above the waves. In the distance, Reid could see a reassuring sight—what looked like the *Solent* riding her anchor and still afloat.

Reid took out a pocketknife and cut a piece of sail to use as a makeshift signal flag which John Askell carried to the very top of the mast. But the *Solent* didn't see them. Possibly from that distance the men simply blended in with the tangle of broken rigging. Reid looked at his watch but it was pointless. The hands of the watch had frozen at 1:25, the moment the *Rhone* sank.

A short time later Reid spotted residents of Salt Island on the beach and tried to get their attention but again, no response. And although the weather was calmer and land was very close, still no one was brave enough to leave the relative safety of the mast and swim for it. It wasn't so much the swelling sea that was intimidating them now.

It was the sharks.

"The sharks were lying around us like little fish," Reese says, and he "didn't dare try and swim to shore."

So the five men resigned themselves to an awkward, restless night. They cut a sail into pieces to use as a shelter and to catch rainwater to quench their thirst. No telling how long they would have to wait for rescue.

Other survivors were still scattered in the sea. Some were together in small groups and some, like John Bailey, entirely alone. The drifting men faced untold hours in the cold and dark. The seas were still high, the rain heavy.

The wind and current were pushing the lucky ones north to land—Beef Island. Ironically, it was the same island where Captain Woolley had intended to ride out the storm.

Although it's hard to imagine, especially after having just witnessed his friend apparently dragged under and killed by a shark, 16-year old John Bailey says as he drifted in open water, he gradually began to feel more relaxed. He was sure by this time that the life buoy would hold him up. Now he just needed to keep his spirits up and he might be alright. He realized how lucky he was. Not only was he able to stay afloat but he had moved far enough from the wreck to be out of danger from the crushing debris.

As darkness began to fall and the seas to settle, John tried to swim toward what he hoped was the direction of land but the effort finally wore him out. With his arms wrapped in the life ring's cords John Bailey rested his head and dropped off to sleep.

Time slipped by until John became aware of a strange feeling—something rough brushing across his leg. Immediately he was jolted awake. Sharks!

No. Not a shark.

He had drifted onto a sandbank. He was close to shore, Beef Island, but not there yet. There was still almost a mile between him and dry land. John Bailey had been in the water now almost 12 hours. He was cold. Freezing. His teeth were chattering and there was an empty feeling in his stomach. But he wasn't hurt. Holding firmly onto his life buoy, John slipped off the sandbar and started paddling to shore, determined to save himself.

It was still dark when John Bailey reached the island. It was a wild place with long thorny grass that "pricks like pins." With no shoes to protect his feet he picked up his life ring and limped along in search of water, food, shelter, other survivors. Was he one of many or the only person to make it? He must have wondered about that as he walked. Then about an hour and a half later, as dawn began to break, he caught sight of two of his fellow shipmates—37-year old yeoman Francis Brewer and 16-year old engineer's servant Fred Pearse. Both had drifted to shore on a plank. Fred was only slightly better off than John having salvaged at least one boot but he was nursing a smashed finger.

The three together now continued searching and soon they came across three local islanders (a black man and two boys) who had heard about the wreck and were headed for the beach to see what if anything worthwhile had washed up. They had no water but they had information. Two other men like them were just down the road. John and the others set off to find them. It was *Rhone* seamen John Palmer and Samuel McCane. They had made it ashore in life buoys. So would another seaman. That meant of the seven buoys on the *Rhone*, at least four had saved lives. Bailey was still carrying his with him. It was his lucky charm and he had no intention of parting with it. Palmer and McCane, however, had no such sentimental feelings. They grumbled that their life buoys had become so sodden with water they nearly dragged them down. Still, they had proved to be better than nothing.

Now they were five. Walking further they spotted smoke from a campfire and met another local man who volunteered to go over to the governor's house about two miles away to get help. For the first time, the shipwrecked sailors felt they could rest. They found oranges and pineapples that washed up on the beach and ate them and before long the governor arrived with a big jug of coffee and some bread.

In the middle of this breakfast feast, six more of the *Rhone*'s crew appeared: George Holdeman (who had been one of the last to see Captain Woolley alive) boatswain's mate Charles Moody, seaman Frederick Hamsford, coal trimmer Thomas Ingram and firemen Henry Buckle and John Jones. And they had quite a story of their own to tell.

As the *Rhone* was sinking, five of the men had managed to grab on to one of the more seaworthy pieces of debris—a large bin used to store seamen's hammocks—and climbed into. It was an odd life raft but at least it seemed to float. They tried, but couldn't steer it and were constantly afraid the thing would capsize. That would have meant death since most of them couldn't swim. Then someone noticed a problem—a hole punched right through the side of the bin. It would surely sink, they thought. And it would have if not for the extraordinary efforts of one man, boatswain's mate Charles Moody. He was nearby in the water and seeing the problem came up with a spar and managed to stick it through the hole to plug the leak. Then holding the pole with one arm and swimming with the other, Moody began towing the bin in the direction they were drifting, toward Beef Island. At about 11 p.m. the bin struck ground, collapsing to pieces. It had gotten them this far but it was not far enough. Like John Bailey, they had come up on a sandbar short of the actual island which was still about a half mile away. Then Moody did something else extraordinary. One by one he took hold of his fellow crew members and swam with them to shore. They were all practically naked—a testament to the trauma they'd been through.

Moody had rescued all but one man, the 19-year old coal trimmer Thomas Ingram, and was about to swim out for the last time to get him when he became confused. Where exactly was Ingram? It was too dark to see. They called out his name. Ingram! No answer. Had he drowned?

Daylight would bring the answer. Ingram was okay. He had heard his crew mates calling but was worried Moody was exhausted and didn't want to risk his friend's life by asking him to make yet another rescue swim. So Ingram had kept quiet and spent the night, uncomfortable but safe, on the sandbar. Remarkably, thanks to Moody, the hero of the hammock bin, all six men had made it.

There were even more survivors. Charles Berry and three others who had been clinging to that overturned boat, also reached Beef Island during the night at about 9 p.m. So had young John Minns. Seaman George Wallace also washed ashore.

The only scheduled passenger to survive the sinking of the *Rhone* was a second-class passenger identified only as an "Italian gentleman from Pennsylvania." He washed ashore in the early evening onto tiny Buck Island a short distance from Beef Island. There is precious little detail about him even from the boatswain, Emmanuel Sherwood, who became his castaway companion. Sherwood was one of the first to come ashore, drifting onto Buck Island on the planks of a "transferring stage" he used as a raft.

Finding each other on the beach, Sherwood says he and

the passenger scrambled together over a small hill littered with prickly pears and had to stop frequently to pull the painful thorns out of their shoeless feet.

After some time the Italian man stopped and declaring, "I will have a pair of shoes," tore up his shirt to make into sandals. They spent an uncomfortable night on the floor of the home of a native couple who were lucky enough to have a house still standing, though it was roofless from the storm.

In the morning the two men linked up with the other survivors, and all were brought by a mail-boat to the *Tyne* which had left St. Thomas an hour after the *Solent* and had managed to ride out the storm in the lee of an island about five miles from Peter Island (possibly Flannigan Island) with relatively little damage.

So the *Solent* and the *Tyne* were accounted for. And Sherwood, looking out over the hills of Tortola, could see the other Royal Mail ship which had been with the *Rhone* at the beginning of the storm, the *Conway*. Captain Hammock had managed, through luck or skill, to keep her from sinking. She had grounded on a relatively sandy shore in Road Town—stripped of her funnel, masts and boats—but had suffered no loss of life.

Ironically, of the four Royal Mail ships near Peter Island that morning, the only ship to sink was the one considered the best of them all—the magnificent *Rhone*. So much had gone wrong. Captain Woolley had miscalculated the storm, lost the *Rhone's* anchor and the ship had hit Salt Island in one of the worst possible spots. If the *Rhone* had drifted 50 yards either way instead of hitting Black Rock dead on, she would have grounded on more forgiving sandy shores.

The *Rhone* might have survived.

Among the last to be rescued were those closest to the wreck of the *Rhone*. John Reid and four others spent almost

20 hours clinging to the mast until sometime after dawn when locals from Salt island who had wandered down to the beach looking to salvage linens, clothes or whatever else might have washed up from the wreck, finally noticed Reid hailing them. Three men rowed over in a punt and took the survivors off the mast in two groups, which was all the little boat could handle.

Once on shore they were greeted by Mr. Abbot, governor of Salt Island. Reid found it a primitive place where people made their living, as the name implies, harvesting salt crystals from the edge of a large shallow pond. They had suffered too in the storm, with every house being knocked down except the governor's. Abbot told Reid that three bodies had washed ashore on the other side of the island. While David Reese borrowed a boat and headed for the *Tyne* for help, Reid and Askell went together to locate the remains of their shipmates.

"The first body we came to was the chief officer of the *Rhone*, Mr. D. Topper," says Reid. Topper had been the first victim of the storm, struck by a spar even before the *Rhone* hit Black Rock Point. The second body, Reid says, belonged to a fireman he remembered playing draughts with. And the third was the *Rhone*'s 27-year old barman Samuel Daly. Reid found him lying on the beach, foam around his mouth. Reid covered them all with bits of cloth and towels that had washed ashore along with some flags. Later George Holdeman says he removed Dalby Topper's ring to bring home to his family. Officers from the *Tyne* came ashore to bury the bodies.

Two witnesses, Holdeman and Whitmarsh, say the body of the *Rhone*'s surgeon, Dr. Richard White, was also buried on Salt Island.

Locals referred to the men they helped bury as "survivors"—people whose lives had been taken by the sea but whose bodies had been given back for a proper burial. For most of the crew and the passengers, including banker Arthur Gibbon and his wife Maria, James and Sophia King, Robert Scott, Henry Arthur Sanderson and Benjamin Walrond, the *Rhone* herself would mark their graves.

As she would for Frederick Woolley. Although John Reid would find his uniform coat, the body of Captain Woolley was never recovered.

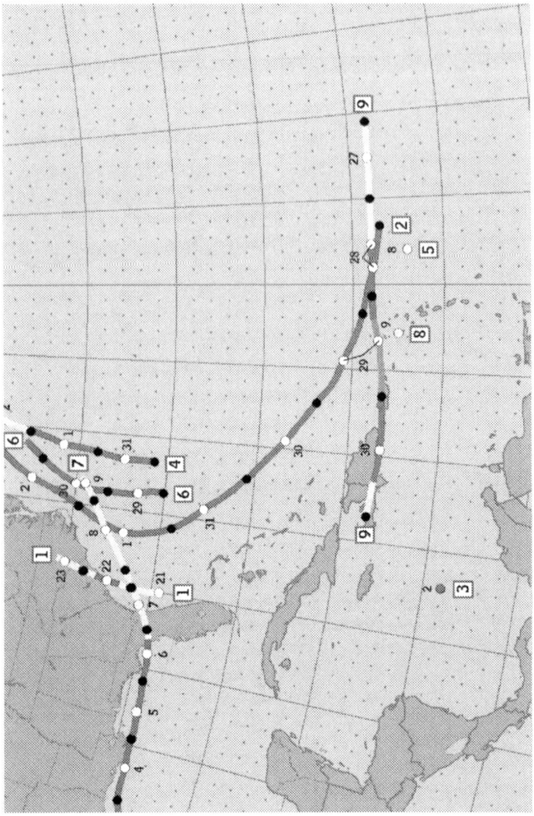

The east to west path of the October 1867 hurricane noted in this graph as Hurricane #9. The path of the storm turned slightly south as it hit the Virgin Islands. Courtesy: NOAA

10

A RIOT OF WRECKS

October 30, 1867
Charlotte Amalie, St. Thomas

Among the first vessels to arrive at St. Thomas after the hurricane was the *Rhone*'s nearly identical sister ship, the *Douro*. The *Douro* had been expected earlier but unfavorable northeasterly winds and a heavy sea had set her behind schedule. When the storm hit the Virgin Islands on October 29, she was still more than two hundred miles away.

Some of the *Douro*'s crew did notice strange weather signs that day—ominous skies to the west, the normally brilliant blue Atlantic turning a dull black. Clearly a big storm was over the horizon but no one seemed to think much of it.

By the next morning, October 30, the weather was glorious and with full sails up, the *Douro* found herself cruising in what one officer called delightful seas toward St. Thomas. As the crew passed within view of Salt Island some thought they saw the dark angles of recent shipwreck but the sunlight reflecting off the glistening waters was so blindingly bright they couldn't be sure what it was. No one imagined on that beautiful day that the indistinct form in the distance was, in fact, the humbled foremast of their sister ship, the *Rhone*.

The afternoon sun was beginning its downward slide when the *Douro* approached Charlotte Amalie, St. Thomas. From about two miles out, the town still appeared quite normal, "very peaceful and very beautiful," according to one officer. But as they got closer "first impressions underwent a sad change."

In the harbor, instead of the iconic forest of masts, the *Douro's* crew encountered a riot of wrecks—more than 60 steamers and sailing ships in all.

Some ships were oddly aground. Others were half submerged with their masts and rigging twisted and tangled. Up on a beach to the left of town were five steamers including the Royal Mail *Derwent*. The large ships were piled one upon another making it hard to tell what funnel belonged to which ship.

The town was equally in ruins. Charlotte Amalie looked as if an army of cannons had razed it to the ground. Houses were roofless. Trees uprooted. The elegant coconut palms which had graced the streets of the town were stripped naked. Accusingly, they pointed upward toward the deceptively dazzling blue sky.

Nearly every business, government office and home had suffered damage. Streets were blocked with all kinds of debris—branches, broken rafters, tiles, carts, slices of zinc roofs, an anchor, tables and chairs. Slabs of marble which had spun around in the air like paper at the height of the storm littered the road. Although the company's agent in St. Thomas, J.B. Cameron, had been in transit and personally escaped the ravages of the storm, his house hadn't. His entire dining room had been lifted up and deposited into the garden of one of his neighbors.

Wharves that used to line the shore were gone. The new floating dock that was within weeks of completion was damaged, rammed by two of the ships that had been anchored in the harbor. The Royal Mail jetties and coaling wharf were also completely destroyed, the coal sheds unroofed and the engineering factory severely damaged.

And hundreds of people were dead. The hurricane had come on so quickly it had caught the busy people of St. Thomas completely by surprise. They'd been casually going about their morning routines—shopping, cleaning, tending to the children—when at about noon a sudden darkness descended on the town. As the rain began to pour down everyone ran for cover. The wind blew so forcefully some had to pull themselves along the streets grasping at door handles and lampposts.

At about 1 p.m., as the skies cleared, people came out by the thousands to inspect the damage not realizing that this was not a typical storm—it was a hurricane. The lull lasted for less than a half-hour before the second part hit sending the terrified townspeople again scrambling for shelter. Many more died.

By the time the *Douro* arrived, almost three hundred bodies, some partially devoured by sharks, had been recovered. Under the guidance of local police, they were being systematically carried away in rough coffins for burial. To retrieve bodies from the wrecks in the harbor, local officials had authorized borrowing diving dresses and hoses from the island of Antigua since the diving equipment in St. Thomas had all been lost.

Of all the shipping companies operating out of St.

Thomas, the Royal Mail Steam Packet Company had suffered the most, with three ships damaged (*Tyne, Conway, Solent*) and three ships wrecked (*Rhone, Derwent, Wye*).

Like the *Rhone,* the *Wye* had been heading out to sea when the storm hit. Her chief officer had given the order to steam away hoping to avoid a collision with other ships in the harbor. But that had proved the wrong thing to do. The *Wye* crashed on the rocks of tiny Buck Island just outside the harbor, broke apart and sank killing some 40 of the ship's officers and crew. The captain of the *Wye* was spared because, as chance would have it, he was on the *Tyne* seeing off a passenger planning to return to Europe on the *Rhone.*

Of all the ships from nine different countries sunk or damaged in the hurricane, by far the greatest loss of life had occurred on the *Rhone.* The most reliable estimate —120 dead.

The exhausted survivors who had been gathered up from Tortola, Beef, Buck and Salt Island, arrived in St Thomas on the *Tyne*. One officer described the shipwrecked men as ghostlike, still wearing their storm-shredded clothes like battle scars, sadly pacing the deck as if walking in a dream.

The plan was for the *Douro* to restock, refuel and return with the survivors to England as quickly as possible.

11

SAD INTELLIGENCE

November 25, 1867
Southampton, England

On a chilly Monday afternoon, a troubled Joseph Stebbing walked along the cobblestone streets of Southampton heading for the Bargate, the medieval tower that was part of the original fortifications of the Old Town.

Stepping through a narrow doorway on the Bargate's southeast corner, he climbed the twenty-six stone steps leading to the upper chamber and turned left into the Guildhall, Southampton's historic meeting place, with its vaulted wood ceilings and slender leaded glass windows overlooking the tailors, perfumers, watchmakers and elegant hotels of busy High Street. The room was packed with political officials, businessmen, clergy and townspeople all wearing similar expressions of deep concern.

This should have been a triumphant moment for Stebbing. A successful entrepreneur from a family of instrument makers and seamen, he had recently been elected mayor of this town he loved so well. But there had been no time to celebrate.

Since early November the town had been reeling from news of a catastrophe—a hurricane in the Caribbean had dealt a deadly blow to Southampton's Royal Mail Steam Packet Company's fleet of ships.

Just a year earlier it would have taken at least two weeks for news of the disaster to reach Europe. That was as fast as a ship could travel. But families in Southampton would learn about the *Rhone* much sooner than that. The world of communication had changed dramatically in the previous several months thanks to what the *New York Herald* called "the grandest work which has ever been attempted by the genius and enterprise of man"—the transatlantic telegraph cable.

Telegraph technology itself was nothing new. For decades lines strung across mountains, hills and prairies had transmitted messages in the dots and dashes of Morse code over thousands of miles. But the signal always ended where the land did.

That was until submarine cables were laid, first in 1851 across the Channel between England and France and then four years later under the Mediterranean Sea. In July 1866, the first reliable transatlantic cable was laid from Valentia Island in Ireland to the town of Heart's Content in Newfoundland, Canada. It could transmit, what was then, an astonishing eight words a minute across the Atlantic Ocean.

Despite the new cable, it would still take several days for the first telegram about the great Virgin Islands hurricane of October 29, 1867, to reach England. Sending a message from New York to London was one thing but first the message had to get to New York. And in the case of the *Rhone* it would happen in stages.

First, on November 1, the French vessel *Panama* left St. Thomas bearing news about the disaster to the outside world. Eventually the *Panama* arrived in Havana where a telegraph cable had been laid just a few weeks earlier connecting Cuba to Key West, Florida. On November 5, that cable carried word of the hurricane from Cuba to the American continent. From there it was wired to New York and from New York across the transatlantic cable to the London Headquarters of the Royal Mail Steam Packet Company.

Finally, on Thursday afternoon, November 7, the "sad intelligence" reached Southampton. *The Times* reprinted text for its readers from some of the original telegrams:

> *"Hurricane at St Thomas. Five steamers and 50 ships ashore."*
>
> *"Fearful hurricane at St. Thomas on the 29th of October."*
>
> *"The Rhone lost at Peter Island, the Wye lost at Buck Island, the Conway Ashore at St Thomas, the Tyne and Solent dismasted but serviceable"*
>
> *"Hurricane worse than '37. Rhone lost, with Woolley and crew."*

It was unfathomable. So many mail packet steamers damaged in one storm. And the best of them all, the *Rhone,* lost with everyone on board! The news spread around town "with the rapidity of lightning."

Early telegraph messages were a lot like today's 140 character bursts on Twitter—information in shorthand that was immediate but incomplete and sometimes confusing or just plain wrong. In this case, several newspapers also reported that the island of Tortola had disappeared entirely under the waves:

> *"The island submerged and all its inhabitants drowned."*

> *"10,000 lives lost."*

> *"We are assured that the island of Tortola has disappeared beneath the sea. Some accounts say it was submerged for eight hours, and that all the inhabitants perished."*

It was, one newspaper said, "one of the most appalling casualties that have occurred in the history of the world."

It was also untrue.

Corrections followed, which raised the hopes of those who didn't want to believe the reports about the *Rhone*. Maybe that was wrong too. As the days passed, more telegrams arrived. Rumors drifted through Southampton like dark clouds. The town waited in dreadful suspense.

On November 16, more than two weeks after the hurricane, there were still no reliable details. And when the *Hampshire Advertiser* printed a list of the officers and crew of the *Rhone*, many feared it to be not just a crew list, but also a death list.

"The people who had friends on board the *Rhone*," reported the *Southampton Times*, "have made up their minds that the worst has happened, and the scenes of distress are most harrowing."

Day after day, bracing themselves against the bitter cold and wintery wind, women gathered at the docks and the nearby offices of the Royal Mail company at #7-8 Canute Road hoping for new information. What had happened? Who was alive and who was dead? When would they know?

Were they wives or were they widows? Should they mourn or could they dare to hope? Southampton was in a state of emotional purgatory.

Any scrap of news was hungrily consumed. At one point a rumor circulated that it had all been a mistake. The *Rhone* had not been lost after all and had, in fact, just arrived in Plymouth. But it wasn't the Royal Mail Steamer *Rhone*. It was another ship also named *Rhone* bound from Amsterdam to Genoa which had docked at Plymouth.

The truth would have to wait until the survivors returned on the *Douro*, perhaps the most anticipated arrival of any ship in Southampton memory. Southampton's Mayor Stebbing decided to delay a public fundraising meeting at the Guildhall in the Bargate until then. No one was sure whose faces they would see when the *Douro* finally docked. *The Times* could not confirm the names and "in the absence of correct and definite information" it announced it would refrain from "adding to the existing excitement by attempting any statement on the point."

Finally, on November 21, the *Douro* arrived—late but safe—back in English waters. She had had a rough crossing, struggling through days of strong northeasterly winds and heavy seas.

That morning the Royal Mail Company announced the *Douro* had reached Plymouth and ordered the company flag hoisted outside its offices to signal there was news. Copies of the telegram were posted at the various newspaper offices. Word spread quickly and anxious groups gathered, grimly waiting for more details.

At 2 p.m. the company posted a list with the names of those who had survived outside its doors. Hundreds gathered around craning their necks to get a look, nervously scanning for names of loved ones. Most searched in vain.

"The people of Southampton," the *Southampton Times* reported, "have for the last fortnight, been dwelling under the shadow of a heavy and almost impenetrable calamity, and though the 'break' has come at last, we regret to say that there is very little of silver lining in the cloud."

Of the reported 145 passengers and crew on the *Rhone*, just 25 had survived.

For some there would have been relief, like John Bailey's parents Henry and Sarah who had waited nearly two weeks hoping against hope for good news. They would soon be reunited with their son. But most were crushed. Carrying with them the unbearable disappointment, they turned away and headed home.

The *Douro* had still not docked at Southampton. She was moored at a buoy about two miles outside of town. The plan was to have a tug rendezvous with her to bring off the mails, passengers and, most importantly, the survivors.

Among the reporters there to cover the story was investigative journalist James Greenwood. The *Morning Star* dispatched this "Amateur Casual" as he sometimes referred to himself, to Southampton as its special correspondent.

With a characteristically skeptical tone, Greenwood pronounced it "stupid" that under the circumstances the *Douro* should have to touch first at Plymouth to deliver the mail, going hours out of her way when the whole world was waiting to hear what the survivors had to say. Then again, he thought, keeping the *Douro* out of sight might be the best way to avoid creating a mob scene at the docks.

On November 22, in a night fog, a tug pulled up to the *Douro* and brought the survivors to the Southampton docks. To Greenwood's eye the shipwrecked mariners looked pretty good, even jolly although oddly dressed in a combination of rags, nautical clothing and costumes donated by people on the *Douro*.

There were two men who appeared particularly interesting. One was John Metcalf. He was dirty, bearded and miserable looking. The other interesting character was the ship's boy, John Bailey. He was young, handsome and impossible to miss because of what he carried around his neck—a large life ring with block letters spelling out the name of the ship that was now a shadowy memory: *RHONE*.

Greenwood wanted to interview both men but he'd have to wait. As soon as they landed they were surrounded by a loud and loving mob. The largest crowds were around John Bailey. Pushing and shoving, they thrust sixpence and pennies at him, wanting to shake his hand and hear his story. Realizing he couldn't compete with the crowd, (in fact, he thought he'd be risking his life if he tried to snatch him from them at that moment) Greenwood quietly made plans to meet later with Bailey.

Not everyone was so pleased to see the survivors. Greenwood says Captain Bevis of the *Douro* seemed to think the honorable thing for all of them, especially John Bailey, would have been to go down with the ship.

"He shook his thick old head," writes Greenwood, "and regarded Master Bailey as if he thought, that, for the credit of a favourite nautical delusion, it would have been more becoming in him to let the buoy alone, and sunk without any fuss."

Greenwood finally got his chance to sit down with John Bailey and interview him over dinner. He found the young man "intelligent looking" with a "respectful manner" and "marvelously cool."

John, of course, had brought his precious life buoy with him. "With it resting against his chair so as to be handy for applying a practical illustration to that part of his narrative it referred to when he came to it, he began."

"We were steaming full head-on and then all of a sudden she went stern on to a rock just close to Salt Island and sort of canted over," John said, going on to describe the darkness, the hail, rain and wind. People were "shouting and shrieking." The surf was pounding on the rocks "throwing up a great sheet all white as snow."

He explained how he had tried to jump off the crippled ship but missed the rocks and landed in the sea, how after his friend was apparently eaten by a shark right before his eyes he had grabbed the life ring and drifted in the night to Beef Island where he reunited with other survivors.

When he ended his story John Bailey seemed satisfied. He "respectfully pulled his hair and backed his chair to notify the fact."

After such an ordeal would he ever go to sea again, Greenwood asked.

"I hope so sir," John replied. "It's no use unless you follow it up, if you make a calling of it."

Greenwood also hoped to talk to John Metcalf but the strange seaman had eluded him. Instead Metcalf strolled into the offices of a competing newspaper, gave them his story and then accepted an invitation to dine with none other than Mayor Stebbing. Later that night, totally by chance, Greenwood did meet Metcalf on the train to London. Everything about Metcalf seemed odd—his ragged mustache and beard, his long untidy hair which was "grotesquely" topped with a Glengarry that one of the passengers on the *Douro* must have given him. Metcalf looked miserable, his face "haggard and lean" as he smoked a dirty little pipe which he must have known was against company rules. There was "rum in John's eyes." More than drunk, John was angry. His name was not on the official list of survivors. Since he was traveling as a "sailor passenger" he was apparently not considered crew and there never was an accurate list of passengers.

"Don't mind me, you know, sir," Metcalf told Greenwood trying to look happy. "I'm nobody, that's what I am—I'm nobody. I was floating around the ocean on a foretopsail yard from two o'clock one day till eight o'clock the next but still I'm nobody; if you don't believe me, look at the list."

Being a "nobody" had become more pressing to his mind than the shipwrecking ordeal he'd been through. At least he was pretty sure his wife wasn't worried. Apparently, no one in his family knew he was still in the merchant service.

In the coming days the appeal of the boy survivor, John Bailey, did not go unnoticed by local dignitaries. One couple in particular took interest in him—Sir Henry James and his wife Lady Anne. James was Director-General of the Ordnance Survey and an enthusiastic proponent of photography as a tool to accurately preserve history. He took some of the very first photographs of the Holy City of Jerusalem and the mysterious stone temple at Stonehenge.

His next project of historic significance would be John Bailey. On his wife's suggestion, James commissioned six photos to be taken of Bailey including some reenacting scenes from John's rescue on the island. It was announced in the Guildhall on November 25 at the West India Hurricane Relief Fund meeting that the postcards would be offered for sale with proceeds benefiting the fund.

The photographs survive to this day—archived at the Royal Society in London. Conspicuously present in every picture is John's precious life ring marked with the name *RHONE*. In two of the photos John is barefoot and scruffily dressed in a well-worn collarless shirt and trousers. Two photos have been embellished with drawings, one portraying John on a tropical beach sitting on his life ring and the other in bed being fed by two native Tortolans, the life ring at his feet. In the final two photos, John is all cleaned up and seated formally—hair combed and parted on the left, dressed in a three-piece suit, tie, shoes and hat in hand. In one of these portraits John Bailey poses with another young survivor, John Minns, who like John is dressed up in suit and tie.

The sale of the Bailey postcards would bring in almost a hundred pounds, adding to the thousands raised for the Relief Fund. The families of those lost from the *Rhone* and the *Wye* were to be compensated from the fund based on salary and circumstance. Careful records were kept of who received what.

Rosina Keeping, for example, the young wife of fireman William Keeping was granted four pounds immediate relief (a sum about equal to her husband's monthly salary) and twelve shillings, six pence per month. She would need the money to take care of herself and her new baby, a son born in February, four months after the sinking of the *Rhone*. Rosina named her son William George after the father and the uncle he would never know.

Rosina would go on to live a long and full life. In 1871, a few weeks before Christmas she married an army man named William Holmes. They moved to Wales and had six more children. Rosina lived to be 87-years old, witnessing the invention of the telephone, the phonograph the electric light bulb and the first skyscrapers.

Not everyone would be as fortunate as Rosina.

Perhaps the most heartbreaking story belongs to the Child family. When Richard Child, the *Rhone*'s third engineer, died he left behind his 42-year old wife Eliza and their ten children. Richard had made a good living earning about ten pounds a month. Without him, Eliza had no money and no job or other way to feed, clothe or house them. The family, according to fund records, lived at 35 St. Mary's Road in Southampton and included six boys and four girls ranging in age from twenty-one to three-years old. Four of the children were under the age of ten. Eliza was allotted the measly sum of five pounds in immediate relief plus twenty-five shillings (just over a pound) per month for some unspecified period of time. But time was not on Eliza's side. In early July 1868, less than a year after the *Rhone* sank, Eliza died. Soon after four of her children (15-year old Ellen, seven-year old Lavinia, five-year old Oscar and three-year old Walter) were sent away to live in orphanages.

As for the surviving crewmen, all received an identical sum—three pounds. It was sufficient to cover the loss of their clothes and that was considered proper compensation.

Besides money, there were also voices raised at the Guildhall meeting that day, commending the *Rhone*'s late Captain Frederick Woolley. Royal Mail Chairman Charles Mangles praised Woolley as man he'd known for years and respected as an experienced officer with a superior record.

MP Russel Gurney also stood to recount to people crowding the hall how he had once sailed with Captain Woolley on a voyage between St. Thomas and England. If there had been any way to save the *Rhone*, Gurney said, he was sure Woolley would have done it. Gurney went on to read from a letter sent by a fellow passenger on that trip vouching for the captain's "skill" and recalling how Woolley "used to express a hope to him that if ever his ship should be lost he should go down in her."

The captain did go down with his ship, into a realm of

legend and wondering. But at least one newspaper, the *Hampshire Advertiser* (whose Chairman was Mayor Joseph Stebbing) raised doubts about whether the loss of the *Rhone* had been unavoidable:

> *Some surprise has been expressed that since Captain Woolley must have had warning of the hurricane he did not seek an offing; but probably something was doing to the machinery which prevented him from putting to sea.*

As it was, Woolley had not read the storm correctly and had not chosen, but was forced, to eventually put out to sea when he lost the anchor. But was the newspaper also alluding to some other mechanical problem on the *Rhone*? Spencer Hall, Woolley's longtime friend, suggests there was something wrong with the *Rhone*, writing in a letter to the *Hampshire Independent*, "It is believed that, had his ship not been at the time undergoing repairs, he might have saved her."

A search of company records, articles and other documents could uncover no other reference to the *Rhone* undergoing repairs around the time she sank.

Hall concludes his letter to the newspaper with a final word defending the character and reputation of his friend. "Of one thing those who knew him will be sure, that thought of his own personal safety would never tempt him to leave the vessel, so long as there was another human being on board."

Although at least one eyewitness account (Reid's) does have Captain Woolley trying to abandon ship before his passengers and crew during those final, fatal moments on the *Rhone*, there is no evidence the company did anything but stand by Woolley calling his performance noble. "Every act of his was guided by a sense of duty and a keen appreciation of the heavy responsibilities of his position," wrote the Acting Consul at St. Thomas, James Lamb. "And by his untimely death the Royal Mail Company have lost the services of a valuable officer, and deep sorrow is brought upon an aged and fond parent."

For the Woolley family, it was yet another painful personal loss. Frederick's mother had now endured the deaths of 17 of her 20 children. Following Frederick's death, she would mentally drift in a kind of mist of memories. Sometimes his mother seemed to forget Frederick was dead "and at other times seems to realize it all," Frederick's sister wrote, "bearing up with the thought that she will soon go to him."

> *It is indeed true our darling Fred has been taken from us in a way we cannot realize. It seems impossible to think that we shall never see his dear sweet face again: it is so awful, so sudden and so unexpected. A few days only before we had the terrible news we had a letter from him saying he had had such a splendid voyage and he was looking forward to his return home with so much pleasure hoping we might all meet again in health and strength. How little we can foresee! Our only consolation is that he was the best fitted to go; and there is the blessing left of feeling that he lived a life of preparation for this hasty and awful summons.*

As for exactly why the *Rhone* sank when the three other Royal Mail ships in the same waters near Peter Island (*Solent, Conway and Tyne*) managed to ride out the storm, Stebbing was leaving that question for others to investigate, and the clergy to explain.

The faithful were instructed to see how God's hand was surely in it.

"There could not be a doubt in the Christian mind," Reverend Dr. Massie told parishioners during Sunday evening services at Kingfield Congressional Church, "that in the recent storm a special providence had been exercised—the lives of some individuals had been preserved amid peculiar dangers. The lad who had floated upon the life-buoy and was cast safely upon the shore, was saved by what had failed to preserve another; the men who were thrown upon the hammocks and were preserved for a long time were saved by no skill of their own whilst others who were placed in circumstances especially favourable to their life were suddenly destroyed, and that without remedy. Thus such storms were under the special control of Him who rules the raging of the sea."

At Portland Chapel, Reverend Charles Williams reassured his congregation "this hurricane, notwithstanding all the losses, deaths, desolations and distress it brought, did not invalidate the doctrine of God's love." God had a reason. Consider the greater good. Just as thunderstorms were necessary to cleanse England's stale summer air, in the Caribbean hurricanes were necessary to make the place habitable. "God had been doing terrible battle, not in enmity but in love, not against His children but for them, against their enemies—fever, pestilence, and death—and for every sailor who has perished in the hurricane, there would be hundreds saved by its ministration from the slower ravages of yellow fever."

"It was the Lord's doing and marvelous in our eyes," declared the Reverend.

It is hard to imagine even the most faithful souls on the *Rhone* walking willingly onto a ship they knew would break to pieces in a hurricane and end up on the bottom of the sea, regardless of how "marvelous" it might on some level be.

But there would be still more to wonder about concerning the mysterious ways of God and nature, because the same islands which had just been struck by a great hurricane were about to experience another awe-inspiring event—a tsunami.

John Bailey. Courtesy:The Royal Society

12

TSUNAMI

Imagine walking from the island of St. Thomas to Tortola without getting your feet wet.

It was possible once.

During the last ice age (about 17,000 years ago) much of the oceans' water was trapped in glaciers and sea levels were about 300 feet lower than they are today. The entire area around the Sir Francis Drake Channel was high and dry. In fact, the Virgin Islands are actually the peaks of an underwater mountain range which sits on a table top plateau about 100 feet below today's sea level.

But this plateau is not sound. Cutting though it like a wound between St. Croix and the main body of the Virgin Islands is a deep fissure more than two miles deep called the Anegada Trough. The chasm is a major cross cutting fault line with the potential to produce powerful earthquakes.

And on November 18, 1867, less than three weeks after a Category 3 hurricane whipped through the islands sinking the *Rhone*, the destructive potential of the Anegada Trough became actual.

Earthquakes in this region are commonplace. The whole Caribbean area is a seismic hot zone honeycombed with fault lines. But this quake would be a historic magnitude 7.5. And it would be just the beginning. Minutes later a tsunami, an unstoppable wall of water estimated in places to be as much as 30 feet high, would crash through the islands. Like the earthquake, it would be one of the biggest ever recorded in this area.

On that day, two ships in particular which had been fortunate enough to escape the hurricane, were unwittingly in harm's way. One was the Royal Mail Steamship *La Plata* which lay at anchor near the southern point of Water Island, about two and a half miles from wreck-chocked St. Thomas harbor.

The other ship was a three-masted, American sloop-of-war called the USS *Monongahela*. Under the command of Commodore S.B. Bissell, the *Monongahela* was in the Virgin Islands on a sensitive and somewhat secretive diplomatic mission. She was to provide transport for American officials who were under orders from US Secretary of State William H. Seward to assist in an upcoming vote on the future of St. Thomas and St. John.

In fact, on October 24, just five few days before the hurricane, the United States had signed an agreement to buy St. Thomas and St. John from Denmark for 7.5 million dollars in gold.

The United States at this time was made up of 37 states and 10 territories in the process of becoming states. William Seward, an ardent expansionist, was in a buying mood. He had recently negotiated the purchase of Alaska from Russia for the bargain price of 7.2 million dollars, just a few cents an acre.

Now Seward turned his sights south. He wanted to establish a naval station on strategically significant St. Thomas so America would have a reliable place in the Caribbean to refuel and repair its ships, something the country had lacked during the American Civil War (1861-1865) when many of the maritime powers including Great Britain had closed West Indian ports to Union shipping. Fortunately, Denmark had sympathized with the Union cause and allowed its navy access to St. Thomas, but the growing nation had no intention of depending on such an arrangement again.

Talks to buy the Danish West Indies began at the end of the Civil War in January 1865, stalled after the assassination of President Abraham Lincoln, then went forward slowly on and off for the next two years until this breakthrough agreement was signed on October 24, 1867. In part because of Denmark's ambivalence, negotiations between the two countries had been kept secret until the last moment. And still it wasn't a done deal.

The treaty required ratification by the U.S. Congress, the Danish Parliament and the people of the two islands who would vote in a plebiscite scheduled for early January. Obtaining the consent of the people was something Seward resisted as time-consuming and unnecessary. It was the Danes who had insisted, calling it "a proper deference to modern European customs."

The *Monongahela* was to be a base of operations for government representatives involved both in coordinating the vote and what was assumed to be imminent transfer of the islands. As his personal emissary, Secretary Seward dispatched the Reverend Charles Hawley to St. Thomas. Hawley's instructions—work toward a "yes" vote but keep your official role quiet. Consider your attendance, wrote Seward "as of a character entirely confidential."

Once in St. Thomas, Hawley supplied Seward with regular reports. The major concern among local businessmen, Hawley wrote, was whether St. Thomas would remain a free port or if the United States would start taxing trade. St. Thomas didn't have agriculture to fall back on. The harbor was everything. That last week of October talk of "the Yankees" and yellow fever dominated local discourse. After more than half a dozen men on the *Monongahela,* fell ill it was decided to move the ship from St. Thomas to St. Croix to take advantage of the island's trade wind breezes.

The 40-mile distance was like being a world away and while in St. Croix the *Monongahela*'s crewmen enjoyed riding parties, dances, picnics, fishing and hunting. On October 25, some of the men were allowed ashore to attend one of the traditional end of hurricane season services of the local churches. On the 29th some heavy weather was noted but it was easily managed with extra anchors.

It wasn't until a few days later that a sailing ship brought word of the disastrous hurricane in St. Thomas and the British Virgin Islands. Commodore Bissell immediately ordered the *Monongahela* to steam back and "render all possible assistance." She covered 36 miles in just three hours arriving at about midnight on November 2.

Among those on the *Monongahela* was Louis Housel, a rookie midshipman and future small-town newspaper publisher. Twenty-four years old with fair hair and gray eyes, Housel (listed as Louis V. Housel in Navy records) was from central Pennsylvania and a recent graduate of Maryland's prestigious Annapolis Naval Academy. At first Housel had been none too pleased with his assignment to the West

Indies. The "most disagreeable station in the world," he called it. But the trip south had changed his opinion. He found himself seduced by the white beaches, green land and bluish electric light called St. Elmo's fire.

Now arriving back at St. Thomas, he was stunned to see the island barely recognizable, the landscape littered with remarkably odd images.

On the beach below Fort Christian, a 40-ton rock had been seized by the wind, lifted up, and a ship's sail spread neatly underneath it like a tablecloth.

A diving bell weighing an estimated nine-tons, had been blown from one place to another, a distance of a quarter mile away.

And a huge beam, which must have flown through the air like an arrow clear across the harbor, had crashed through the roof of a home, piercing first a rocking chair, then a piano before crashing through the floor into the counter of the grocery store below.

Over the following days the sounds of pounding and the clap of hammering filled the air as repairs on this island of storekeepers began. Workmen of whatever specialty became overnight carpenters working at high rates to repair buildings, houses and wharfs damaged by the hurricane. Sunken vessels were raised and sold at public auction. Cargo was salvaged by men in armored diving equipment and advertised for sale at the busy Auction Room.

The storm had put the local St. Thomas newspaper, *Tidende,* temporarily out of business. Its office had been destroyed and five printing presses damaged when the roof collapsed. By using one salvaged press and boxes of new font type recovered from the basement, the editors managed to get a newspaper out on November 13th. With apologies to its readers for missing so many issues, the editors sought to set some things straight.

"None are homeless, naked or starving as some in their mistaken overzeal for us would make it appear," the newspaper reported. "Flour although high in price has not diminished the supply of bread; in fact, all the Bakers, with the exception of one, whose ovens were really injured, produced their bread the morning after the hurricane, and so far from any wish to profit by the emergency of the case, instances can be shown where hundreds of bread were given away to those who seemed for the moment to need them." What had been destroyed in a matter of hours would take years to rebuild. But the community would rebound with "fortitude and resignation."

On November 13, Charles Hawley wrote Seward there was "considerable damage" in St. Thomas but that it "is being rapidly repaired." Despite the catastrophe, negotiations for the sale of the Danish islands to America would continue.

A meeting between American representatives and local officials to discuss the upcoming plebiscite was scheduled for November 18 on St. Croix. The *Monongahela* was to provide transportation for the American commissioners who were apparently more than happy to leave the island. Especially happy to depart was Reverend Hawley. Death was all around. A man in the room above and another next to him had just died, probably from yellow fever. And as they left the hotel, the body of another was being carried down the stairs. Midshipman Housel puts it this way—the commissioners were "desirous of leaving that pestilential 'hole in the wall' to survey the fairer one."

On the evening of Sunday, November 17, as the Royal Mail steamship *Douro* was nearing Southampton with the survivors of the *Rhone*, the *Monongahela* arrived with her passengers at St. Croix's capital, Frederiksted. It was a fine night with temperatures in the high 80s. The next morning dawned peacefully with a smooth sea and warm weather. The delegates left the *Monongahela* and boarded boats headed for their afternoon meeting on the island.

The *Monongahela* was anchored a half-mile off the gently curving new moon-shaped coast of Frederiksted in 60 feet of water. Over her bow to the east, Housel could look directly down the quiet main street of the town. On board the ship, the scene was calm with crewmen reading, writing, or sleeping and others preparing to go ashore on the three o'clock boat.

Suddenly the ship began to shake. It quivered "as if a mighty giant hand laid hold of her and was trying to loosen every timber in her frame," Housel says.

Sailors and officers scrambled up on deck confused about what was happening. The intensity of the shaking and rocking increased and a dreadful buzzing noise surrounded the ship, sounding like a huge swarm of bees descending on them.

"Blowing down the boilers!' said one sailor trying to guess the cause.

But there were no fires lit and Housel knew this.

As the vibrations stopped Housel heard the voice of an old blue-jacketed seaman shouting from the bow with the answer.

"It's an *earthquake*, sir; look ashore!"

Everyone had been so sure something was wrong with the ship they hadn't thought to look much beyond her railings. Now in the distance Housel could see the town enshrouded in a kind of dusty hazy. The stone tower of the English Church was toppled over. All the small buildings had crumbled to the ground. He could make out movement —forms of men, women and children running this way and that. And he could hear a faint noise. It sounded like cries for help.

The far-off spectacle of the tumbled down town would be just the first in a series of unreal sights.

About five minutes after the initial jolt, Housel says, he watched as the island of St. Croix appeared to begin slowly rising right out of the sea. At the same time he heard a new ominous grating sound. Glancing over the bow, he saw the cause. A fierce current was pulling at the ship and, in reply, 60 feet of anchor chain had sprung to life, pulling taut and threatening to tear the ship apart.

Veer chain, the *Monongahela*'s executive officer shouted. Cut the stoppers!

A sailor grabbed an ax and with a great swing struck the chain.

He swung again.

Crack!

Just two blows and the stressed iron anchor chain broke. Like an angry snake, it made a mad race across the deck smashing the carpenter's bench and ladder before making final back somersault in the air and away. Free of the anchor, the bow of the ship rose up in relief.

But now the anchor-less *Monongahela* was free-floating, surging forward and back at the mercy of wild new currents. The ship was 227 feet long and weighed more than two-thousand tons but what happened next would make her seem like a toy boat in a two-year old's bath tub.

The sea swelled. To Housel it seemed as if water materialized from nowhere rising twenty-four feet in seconds. The water rushed into the town, flooding the streets and swallowing homes. The *Monongahela* was being sucked in too, sliding with the currents toward the south end of Frederiksted.

This was unlike anything the officers of the *Monongahela* had trained for at Annapolis. To save the ship they needed to improvise.

Raise the sails!

Catch a breeze they might be able to keep the ship off the town. Several men manned the halyards but the canvas covering was tied tight. Under normal circumstances the heavy knots might be easily undone but by now some of the men were panicking, visibly shaking, and their hands flayed around the ropes uselessly.

Cut it loose, shouted an officer. A sailor grabbed a knife and stabbed at the knots, managing to free a couple of feet of the cover before the knife jumped out of his nervous hands and tumbled overboard.

The order went out to loose the fore topsail. Housel watched as about a dozen men bravely scrambled up the rigging. They had nearly reached the top when they paused, looked out over to the unnatural seascape and came scrambling back down to the deck. Some must have kept their heads, because Commodore Bissell records at this point the sails were successfully set. A light breeze gave him "momentary hope" he might be able to maneuver out to deeper water.

But it was too little and too late. The *Monongahela* had been steadily moving toward the town. She passed over what used to be the beach then, like metal to a magnet, her bow swung north smashing into a frame storehouse and through a row of shade trees.

The strange tour wasn't over because a moment later the water reversed, dragging the ship along with it back toward the ocean. About 500 yards from the edge of town the

Monongahela suddenly stopped. Grounded. Amazed, Housel could now look around and see the exposed bare bottom of the sea floor for a half-mile around him.

And still, this was just the beginning.

Damaging tsunamis are rare and devastating. Among the worst in recorded history were the Japan tsunami in 2011 and the Indian Ocean tsunami in December 2004 which inundated coastal communities in Indonesia, Sri Lanka, India and Thailand killing more than 200,000 people.

Tsunamis form suddenly when a great mass of ocean water is displaced by a landslide, volcanic eruption or an earthquake. Out at sea, a tsunami may be a hundred miles long or more and only a few feet high—barely noticed by ships. But as they approach land and shallower water the wave compresses in length and grows in height becoming a powerful, unstoppable devastating wall of water.

Researchers with the Tsunami Society who scrutinized reports of 91 wave events in the Caribbean dating back to the 1400s found 36 to be verifiable or likely to be true tsunamis. For the others, there was insufficient data or they were attributed to other causes like hurricane storm surges.

But there's no doubt about the wave event of 1867. It was a true tsunami, one of the largest ever in this region.

Aboard the *Monongahela,* Housel and his crew had already witnessed the first phase of the tsunami surge, the drawing back of the water and the returning rush that had carried the ship over the rooftops. But what approached the ship now was nothing short of astonishing.

As far as the eye could see, both to the north and to the south, rose a towering glass-like wall of green water estimated by eyewitness to be between 15 to 30 feet high. Official records call it at more than 22 feet.

For the young midshipman, it seemed as if the all the

water in the town and in the bay had joined forces with the whole of the Atlantic "for a tremendous charge upon us and the shore."

The great wave seemed to pause and hover with dreadful suspense as if gathering strength. The crew was mesmerized. Then it came rushing toward them, massive and unbroken. Many mumbled a prayer and grabbed for anything that might save them from being washed overboard.

Hold fast, came the cries from the men as the wave hit the ship. Housel feared the ship would capsize or he'd be swept away but hardly a drop of water fell on the deck.

Instead, as the wave hit, it slid under the relatively flat-bottomed ship, lifted and carried her rolling side to side, back toward town where she landed—crookedly but upright—on a rocky wharf parallel to shore in front of Robinson's Ship Chandlery Store and Hotel.

The men on the *Monongahela* were safe. But not all of her crew had been on the ship when the tsunami hit. Three were in one of the *Monongahela*'s longboats. Housel had seen them point their boat's bow to face the rising wall of water, rowing hard as they could. They went up and up, nearly topping the inside curve of the wave when their oars seemed to catch in the onslaught of water. The boat slid back and over and was swamped. One of the men would eventually make it to shore but two drowned.

The commodore's coxswain was also killed. He had been loyally manning Bissell's gig even while the *Monongahela* was being dragged to and fro in the crazed current. When the tsunami slammed into the *Monongahela,* the gig was crushed and the coxswain with it.

The shocked men of the *Monongahela* hardly knew what to expect next. When it appeared that the waters were drawing back again to form a new wave, the ship's commander gave the order to abandon ship. At least that's how Housel recalls it.

Interestingly, Commodore Bissell does not mention any such order in his official report to Admiral Palmer. Bissell says his men behaved with "coolness and subordination."

But Housel describes a much more chaotic scene with everyone overcome by the power and mystery of the tsunami. "Every man save himself!" is what Housel heard loud and clear, together with the chief boatswain's mate piping the order with his whistle. Housel, along with many of his fellow crewmen, ran for his life.

Over the side went the ropes. Men scrambled down them like insects and headed in a run for higher ground.

As he was sliding down the rope Housel cut his left hand nearly to the bone, an injury so severe the pain of it would plague him the rest of his life. But at that moment he was running on adrenalin and he didn't pause to notice. He quickly made his way up the flooded streets and, like Lot in the Bible story, he says, didn't dare look back.

The streets were a swamp of debris. Housel waded past chairs, cradles, geese and ducks which flapped their wings in utter confusion. Eventually he caught up with some of his fellow sailors and made it to the foot of a hill where, he says, small groups of whites and blacks were all huddled together crying and praying. Finally, Housel felt safe.

In a frame church on the hill, surgeons worked desperately to try to save the life of another crewman who had landed badly when he jumped from the ship and had broken both his legs. They had to amputate. Doctors operated on the first leg and were about to saw off the other when they realized his internal injuries were too serious.

The sailor was spared suffering through another amputation and later died.

Eventually the commodore would order all men back to the ship and a patrol was sent out to gather up those who had taken advantage of what some called Bissell's previous "kindness of heart."

The watery shock wave had charged across the Caribbean from Puerto Rico to Grenada possibly even reaching the northern coast of South America. The highest wave heights were reported in Guadeloupe.

In St. Thomas, the tsunami would compound the devastation from the hurricane and cause the near destruction of yet another Royal Mail steam ship—the *La Plata*

Like the *Monongahela,* the *La Plata* was caught up in the politics of the day. She had arrived at St. Thomas on November 17 (a day before the tsunami), delivering Special Commissioner Chamberlain Carstensen from the King of Denmark, to help oversee the transfer of the islands to the United States. On the afternoon of the 18th, the *La Plata* was anchored not far from Charlotte Amalie harbor in the process of being coaled by two hulks brought along side. Some of her passengers, bothered by all the noise and the dust, asked to go ashore. Others were curious to see all the ruin caused in town by the recent hurricane so the Royal Mail Superintendent agreed to place a 50-foot steamboat called the *Itchen* at their disposal. Twenty-five people went off to St. Thomas with a plan to return to the *La Plata* at about three o'clock that afternoon on the *Itchen*.

But things did not go as planned. Apparently, the Superintendent didn't trust the *Itchen*'s engineer (a man who had recently been brought in to replace a man drowned during the hurricane) so he decided to appoint a new engineer to the boat and sent Mr. M'Dougal, head engineer of the company's factory, to show him the ropes. As the hour approached 3 p.m.—the time when they should have been picking up the *La Plata*'s passengers in town—the two men were still doing a test run with the *Itchen*.

Just then it started—the rumble of an earthquake. On the *La Plata,* sailors dropped to their knees, says purser William Bayly, and began to scream and pray. About 50 passengers were still on the ship. Some went for boats. Others were calm. Within minutes, Bayly saw the tsunami rushing toward them. People scrambled for the paddle boxes and rigging, the highest points on the ship. Even the bravest seemed to lose hope. *La Plata* was positioned in a bad way, broadside to the monster breaker.

Then at the last minute, caught in the drawback before the rising wave, *La Plata*'s stern swung around. An etching of the event published a few weeks later shows the steamer being lifted up by the approaching wave by the stern, her bow pointing down toward the bottom. The rushing roar was tremendous and water came crashing over the deck into the saloon and cabins. But *La Plata* did not flip. She stayed afloat.

The *Itchen* was not so lucky. As the great roller engulfed the small launch, the *Itchen* turned over killing both Mr. M'Dougal and the engineer. But the loss of life could have been greater. Had the *Itchen* been on schedule and picked up the sightseeing passengers as planned, mostly likely "all would have perished," concluded *The Times.*

Nearby on the USS *Susquehanna*, Rear Admiral J.S. Palmer witnessed the tsunami from the deck of his ship. "The first heavy roller," Palmer noted, "went up into the town swamping the stores, which were mostly on the bay front, floating out, and finally stranding their goods in unheard of directions." Supplies that had been retrieved from the disaster of the previous month were washed away and small vessels were driven up into the street of the town. It was a heartbreaking blow to an already broken town, with all the new damage even "more ruinous to the merchants than that occasioned by the late hurricane."

Admiral Palmer did not remain in St. Thomas long after that. As soon as he got word of the stranding of the USS *Monongahela* he immediately set sail for St. Croix to help with salvaging her.

But this distinguished officer who had had such an illustrious career, who had started in the navy as a ships boy and rose in the ranks to become a Rear Admiral, did not live to see the *Monongahela* re-floated. On December 7, 1867, Palmer died. Cause of death—yellow fever.

To look out over the ocean just after the tsunami hit St. Thomas, it would be hard to tell anything unusual had happened. The sky was cloudless. The bay was still and quiet.

But the impact on the people was profound. Voices raised in churches, town squares and the pages of newspapers wondered why such unforeseen and unpreventable series of disasters had befallen the islands in such a short period of time.

With the Royal Mail offices destroyed, Cameron put an ad in the *Tidende:*

> *WANTED: a suitable place as an office for the Royal Mail Steam Packet Company.*

Once again the *Tidende* found itself apologizing to its readers for the second time in as many weeks. The newspaper was unable to publish its November 20 edition because the temporary offices it had moved to after the hurricane had been wiped out by the earthquake. By Saturday, November 23, the paper was back, calling the recent disaster "the severest Earthquake that (the island) has ever at any previous time been its misfortune to experience."

Were these tragedies, as the St Croix *Avis* put it, a "chastisement from God?" For one religious leader, it was clear the catastrophes served a higher spiritual purpose. It helped fill the pews.

Brother S. Warner Emmanus, disappointed to find his church unusually empty in the days after the hurricane, felt somewhat vindicated after the earthquake. "What one visitation failed to effect, has been done by another of November 18th in a wonderful manner," he wrote. Townspeople were so fearful of aftershocks that they were afraid to go to work and so the church had three meetings every day. "The people now come uncalled to bear the world of God. It is refreshing to see them flocking to the services. May many a sinner be brought to Jesus by these tribulations!"

But perhaps the most lasting effect of these natural disasters would be political—the deal to buy St. Thomas and St. John was foundering. Reverend Hawley, Secretary of State Seward's emissary in the islands, described the effect of the tsunami this way—"This calamity, following so speedily the hurricane of the 29th ultimo, is a serious embarrassment to our mission, as it must, for a time at least, preoccupy public attention."

It's not that people of the islands objected. In January 1868, St. Thomas and St. John (which is only 4 short miles from the island of Tortola) voted overwhelmingly for the sale to the United States. The Danish government also voted in favor and the King Christian IX of Denmark signed the treaty on January 31, 1868.

But now the United States Congress had a serious case of cold feet when it came to expansion in the tropics. Although the treaty was before the Senate, no action was taken by it or by the Foreign Relations Committee. The treaty was shelved with hardly a word of explanation at the time.

It was a complicated situation. Economically, America was struggling and there were questions about the wisdom of the recent multimillion dollar purchase of Alaska from Russia. It could not have helped matters that the Virgin Islands had just suffered through a yellow fever outbreak, Category 3 hurricane, 7.5 magnitude earthquake and a tsunami which had beached one of its own warships—all within a month's time.

Years later, Rear Admiral J.N. Hemphill U.S.N. would confirm that the election commissioners who had been on *Monongahela* told him upon returning to the States, "they were not going to recommend the purchase of hurricanes, earthquakes and yellow fever!" The islands were damaged goods.

Other political storms, including the impeachment trial of the American President Andrew Johnson for violating the Tenure of Office Act, caused more delays. It wasn't until the turn of the century that America seemed ready to try again, signing a new treaty on January 24, 1902. But then, Denmark still resenting its treatment during the failed treaty of 1867, balked.

It would take the threat of war, World War I, to tip the balance. The islands had long been a drain on the Danish economy. And the United States could not afford to have the

West Indies fall under the control of Germany, not with the protection of the Panama Canal at stake.

Finally, on March 31, 1917, more than 50 years after negotiations to buy the islands had begun, the transfer was finalized. In exchange for 25 million dollars the United States purchased all three islands—St. Thomas, St. John and St. Croix—creating the U.S. Virgin Islands.

13

LUCKY AND GOOD

The British Virgin Islands, 1870

On a blue summer day, Captain Robert Woolward cruised past the Virgin Islands, his mood uncharacteristically reflective. Looking out toward the tumble of black boulders on the western tip of Salt Island, Woolward could see the dark angles of what looked like a cross in the water. It was the topmast of the *Rhone*, still visible above the waves It marked the spot where she had sunk three years earlier, a somber reminder of the disaster.

The day the *Rhone* was lost, October 29, 1867, Captain Woolward had been thousands of miles away off the coast of Bahia, Brazil in command of Frederick Woolley's former ship, the *Shannon.* Woolward's log book reflects the details of a relatively uneventful voyage punctuated with the existential realities of life and death at sea.

On October 21 — a 24-year old woman died of fever on board.

On October 26 — a baby boy was born.

Among the *Shannon*'s crew were a dozen lucky men who, like Woolward, had been transferred off the *Rhone* just days before her final voyage. Had they been on the *Rhone,* chances were, they would have gone down with her.

It's also possible that if Robert Woolward had been in command of the *Rhone* for that 10th voyage, the ship and all aboard her could have been spared. Given his experience and skill dealing with hurricanes, Woolward might have realized the October storm was a dangerous hurricane and kept the *Rhone* at the safer harbor at Peter Island instead of trying to weigh anchor and head north as Captain Woolley had.

If Robert Woolward had a close personal relationship with any of the officers or crew lost on the *Rhone*, he doesn't reveal it in his writing. But he does mention at least one survivor, a fireman he calls "Bill."

In what Woolward terms a curious coincidence, a nurse employed by his family happened to be married to this "Bill." When the first telegrams reached Southampton announcing that the *Rhone* had sunk, Woolward's wife Helen broke the news personally to the member of their staff.

The *Rhone* was gone, she told the nurse, and the entire crew feared lost.

"'I knew it," the nurse exclaimed, but not with any sense of shock or sadness. The nurse, in fact, seemed quite cheerful.

"But my Bill is all right, I saw him land in a dream; he went ashore from the wreck on a large, square-looking thing; he's all right.'"

"And so he was," says Robert Woolward. "When the survivors arrived, her Bill was amongst them and he had got ashore on one of the hammock binns!"

So who was this "Bill?" According to official records there was no survivor named William or Bill. However, assuming the name was pseudonym (or possibly a generic name Woolward used because he'd forgotten the firemen's actual name) there are three possibilities: Henry Buckle, John Jones, or Thomas Ingram. All were firemen or coal trimmers on the *Rhone* and all three were saved floating to Beef Island in a hammock bin.

Following the loss of the *Rhone*, Captain Woolward carried on his duties as before—the departures, the arrivals and the storms in between. In fact, passing by the wreck of the *Rhone* on that summer day in 1870, he had only just hours earlier managed to successfully navigate his current ship, the *Neva,* safely around another dangerous hurricane —a storm which nearly sank another vessel which came limping into port with seven feet of water in her hold and two dead.

His *Neva* had come through unscathed and her relieved passengers had gathered with Woolward in the ship's saloon to express their thanks. To mark the occasion one of the passengers read aloud from a letter they had all signed:

> *Dear Sir, as we are now so nearly approaching the end of our transatlantic voyage, we are unwilling to separate without writing to assure you of the feelings which have been called forth by your great judgment, forethought and skill, to which, under the good providence of God, we believe we owe our preservation under circumstances of no ordinary anxiety and danger.*

When it was Woolward's turn to address the passengers, he gave thanks to God for helping deliver them from this "great danger," and thanks also for the knowledge of how to do it. By understanding how hurricanes behaved, Woolward had been able to keep the *Neva* from drifting into the path of that hurricane. But with the wreck of the *Rhone* still in sight, Woolward conceded that in some circumstances knowledge was not enough.

"Once drawn within (a hurricane's) vortex, the best ship became unmanageable," he told them, gesturing toward the tilted sea-lapped mast of the *Rhone*. "Then human efforts were powerless and the finest seamanship was unavailing."

For helping save their lives, the *Neva*'s grateful passengers presented Captain Woolward with a purse containing almost 26 pounds and requested he use the money to buy a silver barometer to mark the occasion and to have it engraved: "From the passengers of the *Neva* to Captain Woolward, in appreciation of his skillful seamanship during a hurricane on the 29th and 30th August 1870."

Not surprisingly, a career as long as Robert Woolward's was not problem-free. In 1882 the notorious rule bender was suspended for two months without pay for the unauthorized "private trading" in turtle.

And he was involved in at least two collisions with other ships for which he was found responsible. In 1874 Woolward was below decks on the *Tagus* putting on warmer clothing against a brisk north wind when he heard the engines reverse. When he came up on deck he noticed a small brig pass close by but was assured by his chief officer the two ships had not touched. After arriving in Southampton, however, the Board of Trade officers informed Captain Woolward the brig had sunk and he was being blamed. The company, in that case, paid compensation for the brig and its cargo of coal.

The whole affair had a curiously positive twist for Woolward. While the inquiry was proceeding, he was temporarily replaced as captain of the *Tagus* by another officer. Shortly after that, the *Tagus* was accidentally hit by another Royal Mail ship and sank. "As had been the case in several other instances, the *Tagus* came to grief soon after I was out of her," says Woodward. Once again, his timing had been fortunate.

Captain Woolward would face a much more serious episode in 1884 when his ship, the *Don*, collided with the sailing yacht *Vanessa* in the English Channel. The *Don* was not seriously damaged but a crewman on the smaller vessel died in the accident. Woolward was brought before a Board of Inquiry which determined he had indeed seen the red light on the starboard bow of the *Vanessa*, somehow lost sight of it and assumed the sail boat had moved off. She hadn't. In this case, regulations clearly called for the steamer *Don* to keep out of the way of the small sailing vessel. Woolward was held responsible but no action was taken against him.

Throughout his career, Woolward managed to personally avoid the major calamities which befell so many of his ships like the *Tagus* and the *Rhone* and the *Rhone*'s twin-sister ship, the *Douro,* which would also sink to the bottom of the sea taking her captain and many others with her.

Built to the very same specifications as the *Rhone* but by a different company, somehow the *Douro* did not inspire Captain Woolward and he spills no ink praising her features as he did the *Rhone*.

Instead he describes long into the night dinner parties on the *Douro* and the amusing frustrations of dealing with particularly difficult passengers including one tiny terror who entertained himself by casting overboard whatever he could get his hands on—hairbrushes, toothbrushes and a large bunch of keys which was needed to unlock the ship's baggage room.

Despite his rank and powers of persuasion there wasn't much Woolward could do to reign in this troublesome young man. He happened to be the grandson of the Brazilian Emperor.

As he had many times before, Robert Woolward would escape the tragic last chapter of this ship as well.

It was the spring of 1882 and the *Douro* was cruising off the coast of Spain under moonlit skies on the last leg of her 62nd voyage. Aboard were 60 passengers, 80 crew and a fortune in gold. In command was Captain Ebenezer Kemp, one of Woolward's former chief officers who he describes as a great officer and a kind man.

At about 10:30 p.m. the *Douro*'s men were positioned on watch—chief officer on the bridge, fourth officer on the quarter-deck aft, an able-bodied seaman at the wheel and two others on lookout.

In the distance the bright light of a Spanish steamer, the *Yrurac Bat*, was spotted— the telltale red light on her starboard bow. Despite this the *Douro* continued full speed for another half hour. It wasn't until the two vessels were almost on top of one another that the *Douro*'s chief officer realized the two ships were actually on a collision course and ordered the *Douro* to be put hard to starboard. But it was too late. The *Yrurac Bat* rammed into the *Douro*'s side, hitting her first amidships just forward of the main mast, then a second time, damaging two boats on the starboard side.

Immediately the *Douro* came to a full stop. Passengers ran out on deck to see what had happened. The effect of collision was devastating—a gaping hole in the ship's starboard side extending well below the water line. Captain Kemp quickly ordered the remaining boats launched and loaded, women and children first as per the renowned unwritten law of the English seaman.

The *Douro* managed to stay afloat for only about twenty minutes. In that time, 54 passengers and 66 crew escaped to safety. But the *Douro* sank before the last boat could be launched. Six passengers and 14 crew including most of the officers and Captain Kemp went down with the ship.

Exactly why the collision had occurred and why the *Douro* sank so quickly were questions taken up three month later in a formal Board of Trade inquiry.

The report praised the crew for their heroic actions following the collision but found the *Douro* entirely responsible for causing the accident in the first place by failing to take "proper steps to keep out of the way" of the *Yrurac Bat.*

And although the *Douro* had eight watertight compartments which should have helped keep her afloat, the board concluded that some of the bulkhead doors were open at the time of the collision. The Board's conclusion:

> *The blow must have damaged the bulkhead separating the after compartment from the engine-room, so that if the doors between the engine and boiler rooms were open, as they probably were, there was nothing to prevent the water from filing all the compartments from the boiler-room aft; and in that case the vessel could not float very long, but would soon go down stern foremost, as we are told that she did.*

Shortly after the *Douro* sank, a person using the pen name "Naval Architect" wrote to *The Times* pointing out what he considered to be serious safety problems throughout the maritime fleet.

The "Naval Architect" said the Board of Trade should require ships to have more and stronger bulkheads which could theoretically make ships unsinkable. “In nearly every steamship,” he wrote, “the mid-ship compartments are of such dimensions that any one of them filled with water would so deeply immerse the hull that the sidelights or portholes would be brought below the water-line and as the bulkheads seldom extend above the main deck there is nothing to prevent the water rising over them and filling the other compartments by the hatch or companionways.”

Passengers were being lulled into a false sense of security and countless lives were being lost unnecessarily, this writer complained, because the government, insurance underwriters and shipping companies were too wedded to tradition and concerned with the bottom line to invest in better safety measures. The editors of *The Times* said they tended to agree with the “Naval Architect.”

The kind of safety improvements the “Naval Architect” called for would be adopted in future ocean liners including on the iconic *Titanic.* In April 1912, when *Titanic* sank on her maiden voyage, despite her 16 watertight compartments, there would again be a rethink on how to design safer ships.

At the time she was lost, the *Douro* was said to be carrying cargo worth hundreds of thousands of dollars—not unusual for a Royal Mail ship returning from Brazil. But the *Douro* went down in very deep water, more than 1000 feet, and whatever treasure she held would lay unseen and untouchable on the ocean floor for more than 100 years.

It wasn’t until 1995, thanks to scientific advances and a determined team lead by Nigel Pickford and Sverker Hallstrom, that the *Douro* was finally salvaged.

Using shipping records, eyewitness reports and sophisticated remote camera equipment Pickford and Hallstrom eventually located what seemed to be the *Douro* on the ocean floor. But they weren’t sure. Certainly it was a very old wreck, but it was nearly unrecognizable as a ship. It was a heap of iron so gutted and shapeless that there was little their underwater cameras could find to positively identify her.

Then they caught a break. A staff member watching the video feed from the site noticed what appeared to be an object on the muddy bottom. It looked like a china plate. Closer inspection revealed that was exactly what it was. And there was a name on the back of the plate—"Royal Mail Steam Packet Company." They had found the *Douro*. And she proved in the end to be quite the treasure ship. Along with the usual wine carafes, bowls, dishes and cups, the salvage company was able to retrieve thousands of gold sovereigns which were later sold at auction.

The loss of the *Douro* would mark a milestone for the Royal Mail Steam Packet Company—the last time (other than wartime) a Royal Mail ship was involved in an accident involving the loss of a passenger.

As for Robert Woolward, he rose to the rank of senior commander in 1892. He and his wife Helen moved to London but Woolward continued to sail Royal Mail ships from Southampton.

By now, however, "Old Woolward" as he was sometimes referred to, was indeed getting old. In 1894, following another incident with the *Don,* Woolward was reprimanded and informed by the company that on account of his age (he was 68) he would have to retire. In recognition of his years of service Woolward was awarded a year's salary—1,000 pounds. Later that same year his memoir, *Nigh on 60 Years at Sea*, was published.

Robert Woolward had been with the Royal Mail Steam Packet Company for a total of 50 years. He had been captain of at least 14 different ships and carried, by his estimate, more than 30,000 passengers ranging from royalty and religious leaders to a lady from York who traveled the world alone with only one handbag and the clothes she had on. In all that time, he never failed to bring his ship safely home to Southampton.

As the saying goes, "I'd rather be lucky than good." Captain Robert Woolward, it seems, was both.

The bowsprit of the Rhone. Courtesy: Dr. Shannon Gore

The Rhone mid-section Courtesy:Dr. Shannon Gore

Swimming through the bow of the Rhone

Author with propeller of the Rhone

14

DIVING THE RHONE

2017

The RMS *Rhone* is still there!

That was the response from a dive operator I contacted four weeks after Hurricane Irma struck the Virgin Islands. Of course, there was little doubt the *Rhone* was still there. Even a category 5 hurricane couldn't sweep away the submerged wreck of a two-thousand ton ship.

Hurricane Irma had made a direct hit on Tortola, Virgin Gorda and their sister islands on September 6, 2017 causing devastating damage to homes, businesses and the basic infrastructure of the Virgin Islands. It would take months, maybe years to recover. But islanders were focused on the future and that included important tourist sites like the wreck of the *Rhone.*

So, the question needed to be asked.

And the answer was that the *Rhone* had been checked out, had no serious damage from the storm and would soon be open for diving. And that, at the time, was good news.

Some have called the *Rhone* the *Titanic* of her day. Although separated in time by almost 50 years, the two ships do share noteworthy similarities.

Both departed, never to return, from the same famous port town of Southampton, England.

Like *Titanic*, the *Rhone* was celebrated as one of the most luxurious ships of her day, a voyager's dream. And if not precisely "unsinkable," it's fair to say during her two years crossing the Atlantic, the *Rhone* had earned a reputation for being one of the safest ships on the sea.

In one respect, however, the *Rhone* is very unlike *Titanic*. While *Titanic* sank in the frigid North Atlantic and lies more than two miles down, so deep it can only be reached by a submersible, the *Rhone* is approachable. She lies in clear, warm, relatively shallow water and is easily visited by snorkelers and recreational scuba divers.

The *Rhone* is not a particularly difficult dive but it's not easy either. The waters around Black Rock Point, Salt Island, where she crashed and sank, can be flat and calm or surging and churning, creating up-swells and currents that fluster some divers. One particular surge pocket in the shallows near her propeller can grab hold of a diver, suddenly sucking them up to the surface.

"We rescue people all the time from the *Rhone*," says Randy Keil, a big-voiced American native who is one of the most respected dive instructors in the area.

But generally, the *Rhone* is considered a safe dive and a popular one attracting thousands of visitors a year. The ship is so big, more than 300 feet long, divers usually explore her in two separate dives.

The bow section of the massive ship rests on its side in about 80 feet of water. The sight of the tip of her gracefully arching clipper ship bow rising from the sand in the deep blue water is eerily beautiful. The bow section is also stable enough to swim through. Over time, the cave-like interior has become carpeted with bright orange cup coral and green and purple sponges. Small schools of yellowtail and red-stripped squirrel-fish wind through the space and lobster often lurk in the darker crevices.

Years ago, a man-sized goliath grouper, named Abraham by the locals, made his home here. The doe-eyed creature was so used to divers he would actually let them stroke him like a family pet. Then in the mid-1980s Abraham vanished. No one knows the details, but it's suspected the gentle giant was speared, quite illegally, by a local fisherman.

In the earliest days of recreational scuba diving, the 1960s and 70s, the *Rhone* was a scuba diver's dream. Dozens of bottles, china plates, silverware, pieces of rigging, port-holes, even a toilet with Wedgwood-blue trim were salvaged from the *Rhone.* Locals say many interesting artifacts were found during the filming of the 1977 underwater Hollywood thriller *The Deep* staring Jacqueline Bisset and Nick Nolte. The *Rhone* served as one of the main sets for the movie and Columbia Pictures technicians placed giant underwater lights all around the *Rhone*, illuminating her interior for the filming. Back then relics were removed, supposedly for 'safe-keeping.' Some were donated to local museums in the BVI but a lot disappeared, were boxed, shelved, sold or otherwise sent off-island to private collections.

"I bet you *Rhone* artifacts are on every continent on the planet except Antarctica," says Clive Petrovic an ecological consultant and longtime diver in the islands.

Clive Petrovic, Bert Kilbride, Michael Decamp, Alan Baskin, Armando Jenik and Randy Keil. They are all names associated with the early days of diving the *Rhone*. But the honor of being the very first person to dive her goes to a man who beat them all by 100 years—hard hat diver Jeremiah Murphy.

Murphy was hired by the Danish government shortly after the hurricane of 1867 to help clear St. Thomas harbor of debris and salvage some of the dozens of ships wrecked or damaged in the storm. To help with this enormous job Jeremiah enlisted his brothers, Andrew and Denis.

An article in the *Port of Spain Gazette* sets the scene. The date is May 1870:

> *We paid a visit to the Murphy's about two weeks ago, they are at Salt Island diving various things out of the wreck of the Steamer RHONE. I have never seen a diving dress, it was a novelty to all of us; the children were delighted to see Murphy in his dress, he certainly cut an awful figure, but when he went overboard and we saw him sinking, sinking until we lost sight of him it was something horrible - the water is 17 fathoms and you can't see the bottom.*

But Jeremiah Murphy could. Through the small glass window in his copper helmet Murphy could view the cavernous hull of the great Royal Mail steamship which had sunk less than three years earlier—her wooden decks, furniture, staircases, large propeller and broken rigging. Her aft mast was still above water at this point, marking the location of the wreck where so many had perished.

Murphy salvaged, among other things, an anchor, chains, copper, a dozen bales of cotton and what was described as a "fine" skull. He also reportedly brought up thousands of dollars worth of valuable bullion and bottles of champagne that were still quite drinkable and were enjoyed in a spontaneous picnic lunch on Salt Island that very afternoon.

1980 marks a major milestone in the history of the *Rhone*. With the support of many of those early divers, the local government declared the wreck site a protected national marine park. Fishing, mooring without a permit and removing artifacts from the area is now strictly prohibited.

Although there are no dishes, sovereigns or other significant artifacts left on the *Rhone* today, there is still plenty to see.

Unlike the bow, the rear part of the *Rhone* is broken open, laying bare what was inside. Exploring the stern is like swimming over a giant puzzle of fascinating pieces. Dotting the iron hull, one can make out hundreds of small rivets which held the vessel together. Near the midsection of the wreck are portholes. Once windows for the passengers to gaze onto dazzling seas, the thick glass is now clouded over and nearly opaque. The inscription on the brass rim of one porthole is readable. It is a number—26.

In the shallowest part of the wreck, about 30 feet below the waves, is the *Rhone's* massive propeller. It is so big, about 18-feet long, divers often don't recognize it and swim right over it assuming it's a piece of the hull.

Following the line of the propeller shaft leads to what is left of the boiler and engine rooms. And further on, sandwiched between hunks of iron, are about a dozen 6" x 6" black and white tiles, most likely part of the galley.

There are still mysteries associated with the wreck of the *Rhone*. The fate of the *Rhone*'s anchors is one. There are none on the wreck site today. Jeremiah Murphy apparently salvaged at least one. George Marler, author of *The Royal Steamer Rhone,* believes he found another—the main anchor of the *Rhone* which was lost in the early hours of the hurricane while she was still at Great Harbor.

But even Marler says he's not one hundred percent certain that it is the *Rhone's* anchor. Local dive master, Sue Thompson, says she and other dive instructors have counted a total of 17 large anchors stuck in the coral on the bottom of Great Harbor. Exactly where was the *Rhone* positioned, that is the question.

"The *Rhone* anchor is one of the biggest down there," she says. "If it is the anchor. You have to reengineer it. It's a cold case."

Another mystery, one that haunts so many shipwrecks around the world and throughout the ages, is what happened to the bodies of those lost in the storm—some 120 men, women and children in all. We know several were buried on Salt Island, others on Tortola but most of the victims were never found. There is, however, an enduring memorial to those who perished on the *Rhone* and the other Royal Mail ship which was in St. Thomas at the time, the *Wye*.

The monument is located in Southampton Old Cemetery, the *Rhone*'s home port. Along with the names of those who died, etched in the fine stonework, is a simple inscription:

> *In Memory of the officers and crews of the Royal Mail Steam Ships RHONE and WYE lost during the hurricane at St Thomas West Indies Oct 29 1867*

History is sometimes lost for a time, but not forever. The *Rhone* lives on for those who choose to remember.

Memorial in Southampton Old Cemetery

AFTERWORD

I approached writing *The Law of Storms* not as a historian (which I am not) but as a journalist. I have worked as a television news magazine producer for more than 20 years, most of that time as an investigative producer for CBS and ABC News. The reporting, research and storytelling skills I developed doing that work were essential for writing this book. Of course, producing a ten minute story for network television is not the same as writing a full length book, but it was an intriguing challenge and I truly loved every moment of it.

Among my first tasks was to construct a timeline. It began with the early days of the Royal Mail Steam Packet Company in 1837 and ended with the discovery of the wreck of the *Douro* in 1995. But the central focus of the timeline was the two years between first and last voyages of the *Rhone* (1865 - 1867) which included the earthquake and subsequent tsunami in the Virgin Islands. I was drawing from so many different sources, each with a different piece of the puzzle, I needed a master document to make the connections and cross-check the different accounts.

From the beginning, I focused my research on finding as many primary sources possible. That would include Royal Mail Steam Packet Company documents, ships log books, port records, government reports, baptism, marriage, birth and death certificates and eyewitness accounts from magazines, newspapers and memoirs published at the time.

Some things I assumed would be easy to figure out, proved surprisingly tricky. There was no contemporaneous summary of the voyages of the *Rhone,* for example. In order to determine the exact number and destinations, I read through every issue of *The Times* of London, the *Hampshire Independent, Southampton Times and* St. Thomas *Tidende* that I could find for the two years the *Rhone* was in operation.

For details concerning the actual sinking of the *Rhone,* I compared the statements of witnesses on the three Royal Mail steamships, including the *Rhone,* that were near Peter Island on the day of the hurricane. I was able to plot, what I believe, is the most complete record of the timing of the storm, the movements and heading of the *Rhone,* the actions and reactions of her crew and captain and how each person who escaped the wreck of the *Rhone* managed to survive that day and night. Among the details that differs from the accounts of others who have written about the *Rhone,* is her final heading as she tried to flee the storm. I rely on three separate eyewitness accounts which state the course of the *Rhone* in that final hour was between Peter Island and tiny Dead Chest Island rather than between Dead Chest and Salt Island where the *Rhone* eventually wrecked.

For details of the sinking of the *London* I relied most heavily on two documents: *The Report of the Official Inquiry ordered by the Board of Trade into the Loss of the Steam Ship London* and the 56-page *Minutes of Evidence of the Official*

Inquiry which includes first hand testimony from 40 experts, witnesses and survivors.

Although this is a work of non-fiction, I hoped it would read more like a novel. To that end, I wanted to hear the characters speak as much as possible. There is no made up dialogue in *The Law of Storms.* The words are taken directly from documents, letters, newspaper and magazine articles in which the individuals are describing what they said or heard others say.

Often to set a scene, I would describe the weather and took my lead from the local newspaper reports.

There were several books that helped put the era in context and provided details on dress, food, culture and history. Among them were *Early History of the British Virgin Islands* by Vernon W. Pickering; *The West Indies and the Spanish Main* by Anthony Trollop; *Macqueen's Legacy: A History of the Royal Mail Line* by Stuart Nicol; *Royal Mail: A Centenary History of the Royal Mail Line 1839-1939* by T. A. Bushell.

The two books that were most essential to me were Captain Robert Woolward's autobiography, *Nigh on Sixty years at Sea* published in 1894 and William Reid's *An Attempt to develop The Law of Storms by means of facts, arranged according to place and time* published in 1838, third edition 1850.

For the purposes of fact-checking and to help identify sources that I might have missed I also read: *The Royal Mail Steamer Rhone* by George and Luana Marler; *Hurricane, Earthquake, Tidal Wave* by Stephen Ashley; *Black Rock and Blue Water* by Andrew C.A. Jampoler and *Twice She Struck* by Michael D. Kent.

My research lead down many dead ends. There are gaps in the historical record. A number of significant RMSPC records were destroyed when London was bombed in World War Two. Other documents were simply discarded by the company years ago. Of course, Captain Frederick Woolley's log book from the *Rhone's* final voyage was lost with her. Sadly, some of the records and relics kept in the Virgin Islands were likely damaged in hurricanes Irma and Maria.

But while researching *The Law of Storms,* more than disappointment, I found myself excited by small discoveries. Among the moments which stand out was finding, early on in my research, the photographs of the ships boy John Bailey. I had read a newspaper article in *The Times* dated December 16, 1867 which mentioned photographs of the "boy Bailey" being taken to raise money for the West India Hurricane Relief Fund and wondered, could those photos still exist? It took some digging but I located them in a collection held by the Royal Society in London and as far as I know I was the first person to request a copy of them in more than a hundred years.

ACKNOWLEDGMENTS

My heartfelt thanks goes out to those in the British Virgin Islands who helped me in my research and indulged my fascination with the RMS *Rhone* over the years including Eva and Alan Baskin, Sue Thompson, Armando Jenik, Randy Keil, Jim Scheiner, Annie and Duncan Muirhead, and more recently Casey and Jeff McNutt.

I owe an inestimable amount of gratitude to my dear friends Roberta Baskin and Jim Trengrove. It was on a holiday trip with them to Tortola in the early 1990s to visit Eva and Roberta's father Alan, that I was first introduced to the BVI and first dove the *Rhone*. Roberta and Jim have been steadily cheering me on as I moved slowly toward the finish line with *The Law of Storms*.

Although it is easier than ever to do research on the internet there are still, thankfully, times when it's necessary to walk into a museum or library and ask for help locating an obscure journal, an archived photograph or for assistance threading the microfiche reader. There are many librarians and researchers whose institutional knowledge and help I very much appreciate including those at the

National Maritime Museum in Greenwich, the Southampton City Archives, the Royal Society, The Enid M. Baa Public Library and Archives in St. Thomas and Widener Library at Harvard University.

Experts in the various topics I attempted to tackle in *The Law of Storms* were generous with their time and insights including Donald Canney and John K. Brown with the ASME History and Heritage Committee.

I was privileged to have met Captain Woolward's great grandchildren, Desmond Woolward and Joy Humphries, who graciously shared family memories and photographs with me.

Writers who have dived into the *Rhone's* story, both before and while I was working on *the Law of Storms,* were generous with their encouragement and ideas including George and Luana Marler, Shephen Ashley and Andrew Jampoler.

I started experimenting with writing *The Law of Storms* while a Nieman Fellow at Harvard University. To my Nieman friends and our writing professor, Rose Moss, thank you.

I am indebted to my dear friend and top-notch journalist Andrea Fleischer and my literature-loving niece Lauren McArdle for reading over the final drafts, catching typos and keeping any awkward phrasing to a minimum. My thanks to other friends who read early drafts and gave me encouragement including Sam Sacks.

And finally, to my husband, Jos Flury. The phrase "supportive spouse" is cliché I know, but I honestly can't imagine having been able to complete this book (and enjoyed it as much as I have) without him. When we came up from our first dive together on the *Rhone* and I announced I thought I could write a book about her story, he was instantly enthusiastic. Later, he even supported my decision to leave a very secure job at ABC News in New York, in part, so that I could spend more time on writing and research. He came with me on road trips to libraries, archives, cemeteries and museums and spent many long hours with me in our home office as I wrote and rewrote the chapters. Jos has been unfailingly encouraging from beginning to end—and that very long part in between.

ABOUT THE AUTHOR

Joan Martelli is an award-winning journalist and television news magazine producer with more than twenty years experience working for ABC, CBS and PBS.

Born in Queens, New York, she is a graduate of Dartmouth College and a Harvard Nieman Fellow. She currently lives in The Netherlands with her husband, Jos Flury. *The Law of Storms* is her first book.

21820046R00117

Printed in Poland
by Amazon Fulfillment
Poland Sp. z o.o., Wrocław